The Chemistry of THE RADIO ELEMENTS

Part 1

MAXWELL PRESS

The Chemistry of THE RADIO ELEMENTS

PART 1

Frederick Soddy, F.R.S.

Winner of Nobel Prize for Chemistry

Lecturer in Physical Chemistry and Radioactivity in the University of Glasgow

MAXWELL PRESS

Chennai Trichy NewDelhi

MAXWELL PRESS

ISBN 978-93-90063-45-1 Maxwell Press

No. 44, Nallathambi Street,
Printed and bound in India Triplicane, Chennai 600 005

MJP 892

Publisher : C. Janarthanan

Publisher's Note

The legacy of a country is in its varied cultural heritage, historical literature, developments in the field of economy and science. The top nations in the world are competing in the field of science, economy and literature. This vast legacy has to be conserved and documented so that it can be bestowed to the future generation. The knowledge of this legacy is slowly getting perished in the present generation due to lack of documentation.

Keeping this in mind, the concern with retrospective acquiring of rare books has been accented recently by the burgeoning reprint industry. Maxwell Press is gratified to retrieve the rare collections with a view to bring back those books that were landmarks in their time.

In this effort, a series of rare books would be republished under the banner, "Maxwell Press". The books in the reprint series have been carefully selected for their contemporary usefulness as well as their historical importance within the intellectual. We reconstruct the book with slight enhancements made for better presentation, without affecting the contents of the original edition.

Most of the works selected for republishing covers a huge range of subjects, from history to anthropology. We believe this reprint edition will be a service to the numerous researchers and practitioners active in this fascinating field. We allow readers to experience the wonder of peering into a scholarly work of the highest order and seminal significance.

MAXWELL PRESS

CONTENTS

THE CHEMISTRY OF THE RADIO-ELEMENTS

GENERAL DESCRIPTION OF RADIO-ACTIVITY

RADIOACTIVITY has introduced into the science of chemistry a new conception. The radioactive elements are radioactive because they are in progress of spontaneous change. The chemistry of the radio-elements is concerned largely with the nature of the products of these changes, their isolation and separate identification.

The property of radioactivity was discovered by M. Henri Becquerel[1] in 1896 for compounds of the element uranium, which he found to be spontaneously emitting new kinds of radiation, very much allied to the X-rays in their general nature. Thus the new radiations pass to varying extent through all matter, quite independently of whether it is opaque or transparent to light. The new radiations, in addition to the properties possessed by light of acting on a photographic plate and of causing certain substances, like the platinocyanides, to fluoresce, resemble the X-rays also in "ionising" the air and other gases, rendering them, for the time being, partial conductors of electricity, and causing them, for example, to discharge a gold leaf electroscope. The pioneer on the chemical side was Mme. Curie, who, working in conjunction with her husband, the late M. Pierre Curie, made the following fundamental generalisation.

Radioactivity is a property of the atom. It is not affected at all by the nature of the chemical combination in which the atom exists, nor by the physical conditions. Up to the present day, in spite of numerous attempts, it has neither been found possible to destroy or diminish the radioactivity of a radio-element by any artificial process nor to cause an element which is not radioactive to become so. The radioactive process goes on at a definite rate and in a definite way which it is at present quite impossible to influence. This, of course, applies to the whole process. In many chemical and physical operations the radioactivity appears to be diminished or removed just as a candle burning in air appears to be destroyed. This is often simply accounted for by the escape during the process of radioactive gases or "emanations," as such have come to be styled. The sum total of radioactivity is not affected by any operation yet tried. In spite of the existence at one time of a vague belief (a belief which has no foundation), that all matter may be to a certain extent radioactive, just as all matter is believed to be to a certain extent magnetic, it is recognised to-day that radioactivity is an exceedingly rare property of matter. Of the eighty or more elements known in 1896, the year of the first discovery of the property, two only, uranium and thorium, are certainly radioactive in the now accepted full sense of that term. These two elements are at the end of the Periodic Table with the heaviest atoms known (238.5 and 232.4), and before them in the Table is a great gap, the next heaviest atoms being those of bismuth (208.0) and lead (207.1) (O = 16). Radioactivity has peopled this gap with thirty new elements. Two other elements, potassium and rubidium, possess a feeble indication of a radioactivity of a peculiar kind, but it is far from proved that this radioactivity is of the same nature as in the case of uranium and thorium, and for the present it need not be further considered. To-day, however, some thirty distinct types of radioactive matter are known with certainty, each with a definite and characteristic kind of radioactivity. Radioactivity is an atomic phenomenon, and each of these new types is fully entitled to be considered a new atom, and therefore a new chemical element. These are all derived from natural minerals containing uranium or thorium, and their isolation and identification as separate individuals is due entirely to the delicacy of radioactive methods, which far surpass in this respect the utmost that the spectroscope can accomplish.

The first discovered of these new radio-elements were radium, polonium,[2] and actinium.[3] They were derived from the working up of large quantities of pitch-blende (impure uranoso-uranic oxide, U_8O_8) from the Joachimsthal mine in Bohemia. Latterly a similar study of the thorium minerals has resulted in the isolation of similar new radio-elements of technical value, mesothorium and radiothorium. In all of these the radioactivity of the material is relatively enormous compared with that of uranium, and is more or less permanent. Of these new types of chemical elements only radium and its first product, the radium emanation, have been thoroughly examined in the same way as the ordinary elements. For these the atomic weight has been directly determined and the spectrum mapped. All the others either exist in too small quantity for this yet to have been possible, or are incapable of being separated from the inactive elements with which they are always associated in the mineral. In radium and actinium no diminution of the radioactivity with time after separation has yet been observed, and the same is true of the common radio-elements, uranium and thorium, while in the others a slow decay over a period of years takes place. Besides these, numerous other types of radioactive matter are known which are more transitory, lasting in the various cases only months, days, hours, minutes, or even seconds.

Some of the earliest of these types of transitorily radioactive matter to be studied were the emanations.[4] The radio-elements, thorium, radium, and actinium, but not uranium or polonium, produce, in addition to radiations, what are known as radioactive emanations, and the two phenomena must on no account be confused. A current of air passed through the first-named substances, preferably in solution, carries away radioactive gases or "emanations" which give out rays of similar kind to those from the other radioactive elements. The radioactivity of the emanations is, however, transient, lasting a few days in the case of radium, a few minutes in the case of thorium, and a few seconds in the case of actinium. The emanations can all be condensed out from the air at the temperature of liquid air. Chemically they resemble the rare gases of the argon group in that it is impossible to absorb them by chemical reagents or cause them to enter into chemical combinations. The emanations all have the power of imparting radioactivity to solid objects coming in contact with them (also called the "induced" or "excited activity").[5] This imparted activity again is transient, decaying according to regular laws at characteristic rates in the three cases. That from thorium lasts a few days, whilst that from actinium and from radium lasts a few hours before the decay is complete, the imparted activity from actinium decaying slightly more slowly than that from radium.

A chemical examination of the radioactivity of thorium revealed the following remarkable facts. Simple precipitation of the thorium from its solution gives thorium hydroxide, from which the major part of the radioactivity is found to have been removed. The α-radiation is much enfeebled, and the substance has completely lost the power of producing the radioactive emanation. The solution from which the thorium had been precipitated possesses the whole of the emanating power of the original solution, and on evaporation and ignition, to expel ammonium salts, gives a minute residue containing the whole of the radioactivity which the thorium has lost. This new radioactive substance separated chemically from thorium was termed thorium X. The activity of the thorium X is, however, only transient. In the course of a month it completely disappears, decaying to half its initial value in four days. The emanating power disappears at the same rate. The thorium hydroxide, however, at first but slightly radioactive, *recovers* its radioactivity and emanating power just as fast as that of the thorium X decays. In a month its activity regains its old value. Now a fresh treatment with ammonia separates anew a fresh amount of "thorium X" of the same activity as the first. Thorium X is a short-lived radioactive product of the thorium, which itself produces the gaseous emanation, and the emanation in its turn produces the active deposit responsible for the "excited" or "induced activity." The case has been treated at length because it was the first "disintegration series" to be elucidated, and is typical of all radioactive changes.[6]

As a result of these and similar observations, a complete theory of radioactivity was put forward which is now universally accepted.[7] The radio-elements are unstable and change spontaneously, the change being accompanied by the expulsion from the original atom of "rays" and the production of a new type of radioactive atom. The latter is often very much more unstable than the parent element, and changes again with expulsion of rays and formation of another new atom. This process of successive changes proceeds often through a great number of steps. Thus an α-ray is expelled in the change producing thorium X, the thorium X atom expels another α-ray and changes into the gaseous emanation, the emanation atom expels another α-ray and changes into the non-volatile matter constituting the active deposit, which in turn undergoes further changes of precisely similar type. The extension of this view has gradually included all the known phenomena of radioactivity. The thirty new types of radioactive matter before mentioned which occur in the uranium and thorium minerals are all almost certainly produced from either uranium or thorium by this process of successive changes. Lacunæ still remain to be filled. The production of radium and of actinium from uranium has not yet been directly observed, but there is strong indirect evidence that it occurs, and that time alone is required for the experiments to yield positive results.

Very definite conclusions have been established as to the precise nature of the atomic disintegration. In spite of very great differences in the stability of the various radio-elements, the periods during which they exist varying from thousandths of a second up to geological epochs of time, all the changes are of the same type. The rate of change varies enormously, but the changes all obey the same law as that of a monomolecular reaction. As expressed by Ostwald in words, this is: "As time increases in arithmetical progression, the quantity of substance decreases in geometrical progression." This is the simplest law possible. The rate of change at every instant is proportional to the quantity remaining unchanged at that instant. It has been established that radioactivity, so far as the individual atom is concerned, is an instantaneous phenomenon. Before and up to the actual moment of disintegration the atom of the radio-element is in no way distinguished from the atom of an inactive element. After the disintegration has occurred, the new atom produced is in no way different from an ordinary atom until it in turn again disintegrates. But in any mass of a radio-element the atoms do not all disintegrate at the same instant, so that the emission of rays appears at first sight to be a continuous phenomenon. The simple resolution by a lens of the glow produced by α-rays on a phosphorescent screen of zinc sulphide, as in the spinthariscope of Sir William Crookes,[8] is sufficient to show that the emission of the radiation is perfectly discontinuous. The rays from the radioactive substances have been proved by physical methods to consist, at least for the most part, of projected particles expelled from the substance radially in all directions with great velocity.[9] This forms an essential part of the theory of atomic disintegration. Three types of radiation are distinguished. They are, in order of penetrating power, the α-, β-, and γ-radiation, the α-rays being very feebly penetrating, but by far the most energetic and important type.

In all cases where the disintegration is accompanied by the type of radiation known as the α-rays, the atom suffers a definite loss of its mass, part of it being projected as a separate atom with great velocity. It was predicted from the association of helium in the minerals containing the radio-elements that one of the ultimate products of the disintegration was the element helium. In 1903 the continuous production of helium from radium was directly observed by spectroscopic methods.[10] The α-rays were shown first to consist of positively charged particles of atomic size expelled with a velocity about one-tenth that of light. The existence of this charge renders the particle liable to deviation by electric and magnetic forces. By the application of these, and similar physical methods of attack, data were gradually accumulated from which the atomic weight of the radiated particle constituting the α-rays could be determined. This proved to have the value 4 in terms of hydrogen as unity. The α-particle carries two atomic charges of positive electricity; that is to say, it is a divalent ion. The identity of the α-particles with the helium produced in radioactivity has recently been established by experiments in thin-walled glass tubes of capillary bore, perfectly gas-tight, but so excessively thin that the α-rays are able to penetrate the walls.[11] Such a tube, filled with the gaseous radium emanation, generates helium in the surrounding space *outside* the tube, though when filled with helium none escapes. If the tube is wrapped round with lead-foil the helium is driven into the lead, and can be so stored. On melting the lead in a perfect vacuum the helium is liberated, and can be detected by its spectrum.

The β-rays also consist of expelled particles, but in this case the particles are not atoms of matter, but atoms of negative electricity or "electrons." The velocity with which the β-particles move varies in different cases from about 25 % to very nearly 100 % of that of light. They have considerable penetrating power, and are able to traverse thin plates of metals and other substances before being completely absorbed.

A third type of radiation, the γ-radiation, usually accompanies the β-radiation. For long it was supposed to be a secondary electro-magnetic disturbance in the ether due to the motion of the β-rays, and the same explanation embraced also the relation between X-rays and cathode-rays. But now opinion is divided as to whether the γ-rays, and possibly the X-rays, are not also radiated particles, consisting of an electrically neutral doublet of one negative electron and one, still itself unknown, positive electron.[12]

The three types are distinguished in the first place by their power of penetrating matter. The α-rays are most easily absorbed. None are able to penetrate a piece of ordinary paper or a few centimetres of air at atmospheric pressure. The β-rays go through thin metal foils with ease, but are for the most part absorbed by a millimetre of lead. The γ-rays are able to pass through great thickness of metal without complete absorption.. The more penetrating types are those of the thorium and radium series, and for these every 1.4 cm. of lead cuts down the radiation to about half its initial value. The three types are distinguished also by their behaviour in the magnetic field. The α- and β-types are deviated, the γ are not. The β-rays, like the cathode-rays or radiant matter of the Crookes tube, being negatively charged and almost mass-less, are readily coiled into circles or spirals by a magnetic field. Some of the

β-rays travel with a velocity very nearly that of light, which is many times greater than that of the cathode-rays, so that the β-rays are considerably more difficult to deviate than the latter. The α-rays, being positively charged, are deviated in the opposite sense to the β-rays, but owing to the mass being much greater than that of the β-rays, the deviation is extremely slight, and refined experiments with very powerful magnetic fields are necessary to detect the deviation.

Recently a great deal of work has been done which goes to show that in addition to the prominent β-rays, of great penetrating power, numerous types of very soft or feebly penetrating β-rays are expelled. Throughout, therefore, two kinds of β-rays have been distinguished. The β-rays of the kind known when the term was adopted are penetrating rays capable of passing without appreciable absorption through thin metal foils which completely absorb the α-radiation. Latterly many of the changes thought not to give rays have been found to give out β-rays of very low velocity, and therefore very easily absorbed, and these can often be distinguished as a specific type of radiation only by elaborate experiments. They are therefore shown in the diagrams in brackets :—thus (β). They cause considerable ionisation over short ranges, and have often been confused with the α-rays. A still slower type of β-rays, even less penetrating than α-rays, and quite incapable of ionising gases, is known by the name of the δ-rays, but these are not considered here. It is probable that they are a secondary radiation set up by the impact of α-rays on matter.

The α-rays have some very remarkable properties, which are unique to this type of radiation. They travel a definite distance, known as the "range," in any homogeneous medium, before being absorbed. The various types of α-rays expelled in the different disintegrations differ from one another *only* in the initial velocity of expulsion, which varies from 7.5 % of the velocity of light for the α-particles of thorium C to about 5 % for the α-particles of uranium. It has been proved that all the α-particles expelled in any one type of disintegration travel with exactly the same velocity.[18] This velocity is gradually diminished to exactly the same extent for each particle in passage through a homogeneous absorbing medium, until the "critical velocity," 2.7 % of that of light, is reached. The ionisation produced, and the same is true generally of the photographic and fluorescent effects, in any given length of its path *increases* as the velocity of the particle diminishes down to the critical velocity, when all effects cease with great abruptness, and the α-particle is "absorbed" or passes beyond all means of detection. Hence the "range" of the α-particle, which is proportional to the third power of its velocity, is an important constant, characteristic of the type of disintegration in which it is expelled. The ranges given for the α-rays of the uranium and actinium series are recent determinations in air at N.T.P.[14] Those of the thorium series are less definite as regards temperature, and are taken for the most part from the tables in *Le Radium*.[15]

A connection exists between the range or velocity of the α-particle and the period of the change in which it is generated. The longer the period the shorter is the range. If the logarithms of the periods are plotted against the logarithms of the ranges, straight lines result in the uranium and actinium series. The thorium series has not yet been worked out.[14] All substances absorb α-rays proportionally to the square root of the atomic weight of the substance, if elementary, or to the sum of the square roots of the weights of the constituent atoms, if a compound or mixture. The range in air at N.T.P. varies between the limits of 8.6 cm. and 2.6 cm., so that knowing this range, the range in any other substance simple or compound can be calculated.[16] In mica the fastest α-particle known travels 0.40 mm. The pleochroic halos occurring in certain micas are due to microscopic inclusions of radioactive substances, the α-rays of which have caused spherical shells of blackening around the inclusion, the ranges of which correspond to the various ranges of the α-rays emitted. From the diameter of these halos the thorium inclusions can readily be distinguished from the uranium inclusions, which are somewhat smaller.[17]

Three main effects produced by all the rays more or less in common are chiefly used in their detection and measurement. The rays act on the photographic plate, cause certain fluorescent substances to glow, and ionise the air and other gases. Photographically the α-rays have slight action, while their ionising action is enormous compared with that produced by the other two types. The β- and γ-rays possess, relatively to the α-rays, rather feeble ionising action, but are chiefly operative in the photographic actions, the action of the γ-rays being always very feeble compared with that of the β-rays which accompany them.

Different fluorescent substances behave differently to the three types of rays. For the α-rays a preparation of zinc sulphide is most sensitive. For the β- and γ-rays, barium and other platinocyanides and willemite (zinc silicate) are the fluorescers usually employed. Magnesium platinocyanide, which responds brilliantly to X-rays, is little affected by β- and γ-rays. The scintillations shown by a lens in the fluorescence of zinc sulphide under the α-rays are due to the individual impacts of the rays, each impact causing a separate flash of light. This is one of the methods by which the actual number of α-particles expelled from a radioactive substance in a given time has been directly counted.[18]

Radioactivity being the manifestation of a spontaneous process of atomic change or transformation, the least amount of any type of radioactive matter that can be detected is the least amount in which this change can be detected. So that if the type changes rapidly, proportionately less of it will give a detectable effect than if it changes slowly. Even for the most slowly changing types, thorium and uranium, the sensitiveness of the radioactive tests are considerably greater than, though of the same general order as, the sensitiveness of the usual chemical reactions of these elements. For the most rapidly changing types, the presence of a few hundred individual atoms is sufficient to give a clear radioactive effect. The significance of this will be the better understood when it is remembered that the most delicate spectroscope test of matter known requires not less than many millions of millions of individual atoms. On this account and because of the evanescent character of the materials the chemistry of the radio-elements has been described as a "Chemistry of Phantoms."

Although the actual quantities of matter producing radioactive effects are usually infinitesimal, they are, however, always capable of being calculated from the magnitude of the radioactive effect and the period of average life of the radio-atom producing the effect. Thus the calculated quantity of polonium in a radium mineral is 1 milligram of polonium for every 14 tons of uranium. It is possible therefore to know beforehand, in almost every case, whether or not sufficient of the material is present to warrant attempts at its separation and the examination of its spectrum and atomic weight. This is indeed very rarely possible, but the subject has been attacked by new methods and much information has already been gained of the chemical nature and reactions of these ephemeral elements. A special branch of chemistry, which may be appropriately termed "Radio-Chemistry," has come into existence, the scope and methods of which are largely peculiar to itself. The results, however, throw new light on many chemical problems.

RADIOACTIVE CONSTANTS, PERIODS OF AVERAGE LIFE AND RADIOACTIVE EQUILIBRIUM

THE most important property of a radio-element is its period of average life or its rate of change. This is not only a natural fundamental time-constant, serving to identify the peculiar element in all circumstances and conversely affording an absolute standard of time throughout the universe, but it fixes also the whole character of the knowledge it is possible to obtain as to the nature of the element. In the case of a rapidly changing substance, the period is determined by direct observation of the rate of decay of the radiation with time. Provided the product can be obtained in the homogeneous state and the products of its disintegration do not interfere, as is, for example, the case for polonium, the radioactivity decays in an exponential manner with the time according to the equation

$$\frac{I_t}{I_o} = \epsilon^{-\lambda t}, \text{ or } \log_{10}\frac{I_t}{I_o} = 0.4343\lambda t,$$

where I_o is the initial activity, I_t the activity after any time t, λ a constant known as the radioactive constant, and ϵ is the number 2.71828. The physical meaning of the radioactive constant λ is the fraction of the total amount of radioactive substance disintegrating in the unit of time (provided that the time unit considered is so small that the quantity at the end of the time unit is not sensibly different from that at the beginning). The time required for one-half the radioactive substance to change is known as the period of half-change T, and is calculated from λ by means of the equation

$$\log 0.5 = 0.4343\lambda T.$$

Hence T is 0.6932 $(1/\lambda)$, or $1/\lambda = 1.443T$.

Now it can be shown mathematically that in a system changing according to the law of radioactive change, the period of average life of the atom is $1/\lambda$. The term "period of average life" relates to the future life of the atom. It has nothing to do with and is not affected by the period the atom has already been in existence. Any time may be chosen as the starting point. The period of average life is the sum of the separate periods of future existence of all the individual atoms divided by the number in existence at the starting point.

There is thus a simple connection between the radioactive constant, the period of average life, and the period of half-change. Each of the latter periods have their conveniences in rapid mental calculation. Suppose it were desired to know the fraction of a mass of radium changing in a time so short that the total mass is not much diminished, say a few years. The period of average life of radium is 2500 years. The radioactive constant or the amount changing per year is 1/2500 (which is usually expressed $\lambda(\text{year})^{-1} = 4 \times 10^{-4}$), and the fraction changing in any time, provided it is not too long, can be found from this by simple proportion. But if it were desired to find the amount of polonium (period of average life 202 days, period of half-change 143 days) remaining unchanged after say three years, the following application of the period of half-change would be useful. In n times the period of half-change the amount of a radioactive substance is reduced to $(\frac{1}{2})^n$. Three years is 7.66 times the period of half-change. Hence the amount of polonium remaining is between $(1/2)^7$ and $(1/2)^8$, *i.e.* between 1/128th and 1/256th, and by the use of logarithms can be found at once to be 0.0492. It is worth remembering that in ten times the period of half-change the quantity remaining is less than 0.1 %, and in twenty times the period of half-change less than a millionth.

Absorption of β- and γ-Rays.—It happens that the foregoing considerations are applicable in a totally different connection in radioactivity, and it will save repetition to refer to them incidentally here. The law of absorption of a homogeneous β- or γ-radiation is mathematically the same as that of radioactive change. The absorption, or diminution in the amount of these radiations, is proportional always to the amount of radiation remaining unabsorbed, and therefore follows the exponential law denoted by the equation

$$\frac{I_t}{I_o} = e^{-\mu t}$$

where I_o and I_t are respectively the intensities of radiation initially and after passage through any thickness t, and μ is the "coefficient of absorption." The radiation is half-absorbed in a thickness T given by $0.693/\mu$. The reciprocal of the absorption coefficient $1/\mu$ is the "average path" of the individual rays, on the theory that absorption is due to the sudden stoppage of individual rays by the absorbing matter, and not to a gradual loss of energy of the whole beam. In the sequel the thickness of aluminium in mm. necessary to absorb half the β-radiation is given in each case, and the thickness of lead or other metals in cm. in the case of the γ-rays. From these the absorption coefficient can be deduced by the above relation, but the thickness of half-absorption is the more generally useful. It must be understood that these laws are not always followed exactly. Some of the β-rays are not homogeneous, notably those of mesothorium 2 and of radium C. The β-radiation of the radium family has recently been analysed into no less than seven distinct groups with velocities ranging from 0.98 to 0.63 that of light.[18A] In the case of the γ-rays the exponential law holds after thicknesses greater than the equivalent of 1 cm. of lead have been traversed for all γ-rays except those of actinium C and probably radium E, which are very much "softer" than the others. The "density law," that the absorption coefficient multiplied by the density of the material is the same for all materials, which holds roughly for most of the newer types of rays, holds most closely for the γ-rays, though here the heavy metals, such as lead and mercury, are often rather exceptional. The numerical value of the absorption coefficient depends also to a considerable extent upon the experimental disposition employed.

Radioactive Equilibrium.—The radioactive disintegration series has been likened to a number of reservoirs of different heights, but of the same cross section, with capillary tubes of different sizes in the bottoms, supplied at one end with a steady flow of water which runs through the whole series, from the bottom of each reservoir into the top of the next. The steady flow of water in at the top represents the steady production of the first disintegration product from the parent. The period of the latter is of the order of thousands of millions of years, so that the diminution of the quantity of the parent is quite inappreciable over short periods, and consequently the supply of the first product is constant. The water flowing out from one reservoir to the next represents the proportion of one member of the series continuously being transformed into the next. The reservoirs with small outlets represent the radio-elements of long-life, the size of the outlet being inversely proportional to the life, or directly proportional to the rate of change. In these reservoirs the water accumulates till a certain "head" or pressure is produced at which the outflow and intake are equal. The law of radioactive change, that the amount transformed in unit time is proportional to the amount present, is the same as the rate of flow of water under the circumstances of the analogy, the amount flowing out being proportional to the amount of water in the reservoir, that is to its height or head. A system of the kind described will, after the water has been running a sufficient time, come into equilibrium, and then the outflow and intake are for each reservoir throughout the series the same. In a radioactive mineral in which the products of transformation accumulate and do not escape, the amount of each product formed is equal to the amount transformed per unit of time throughout the series. Such is known as radioactive equilibrium. But by definition of the radioactive constant λ, if the amounts of the successive members are Q_1, Q_2, Q_3 . . . Q_n, in radioactive equilibrium

$$\lambda_1 Q_1 = \lambda_2 Q_2 = \lambda_3 Q_3 = \ \ldots \ \lambda_n Q_n$$

where λ_1, λ_2, λ_3 . . . λ_n are the respective radioactive constants. Therefore the relative amounts of the successive members of the disintegration series in equilibrium are inversely proportional to their radioactive constants, or directly proportional to their periods of average life. Thus in uranium minerals there is a constant proportion between the quantities of the successive members. There is always about 3,200,000 times as much uranium as radium, that is, there are about 3 tons of uranium per gram of radium, so that if the period of average life of radium is known to be 2500 years, that of uranium is about 8,000,000,000 years. In this way have been determined the periods of the very slowly changing elements on the one hand, and, on the other, the amounts of the very short-lived products in minerals. In the first place the period cannot be directly determined, but the relative quantities can. In the second case the periods can be determined, but the quantities are too small for direct measurement. For the radium emanation, which is a gas, the quantity, though infinitesimal, is capable of exact measurement. The volume of emanation in equilibrium with 1 gram of radium (element) is about 0.6 cub. mm. measured at normal temperature and pressure. The period of average life is known very exactly by direct observation of the rate of decay of its radioactivity. Hence for this one member both data are known, with the result that the same information can be obtained by calculation for most of the other members of the uranium series and for uranium itself. In the case of the thorium and actinium series the periods of the primary elements are less perfectly known. It should be stated that if "branch series" occur these calculations are to a certain extent vitiated. The order of the results is, however, probably correct.

Some very important consequences, most easily illustrated by specific examples, follow from the foregoing considerations. The period fixes not only the rate of decay of the radioactivity of a radio-element after it has been separated from its parent, but it also fixes the rate at which that radio-element is regenerated by its parent. Thus the β-activity of uranium X, when separated from uranium, decays with a half-period of 24.6 days. In the uranium the β-radiation is regenerated to the extent of one-half its normal equilibrium value in 24.6 days. In 49.2 days the β-radiation of the uranium X decays to one-fourth of its initial, while that of the uranium increases to three-fourths of its final value.

The equilibrium quantity, and the amount of radioactivity corresponding to this quantity, form the natural units by which the amount of substance, or of its radioactivity, are measured. In a uranium preparation the quantity of uranium X and the β-activity corresponding thereto increase with time until the equilibrium value is reached, and then do not further change. In seven times the period of half-change the quantity is within 1 %, in ten times within 0.1 %, in thirteen times within not much more than 0.01 % of the equilibrium. It is the period of the product, therefore, which fixes the rate of recovery as well as the rate of decay after equilibrium has been disturbed. A rapidly changing product is regenerated by the parent with corresponding rapidity, a slowly changing product with corresponding slowness. These considerations apply especially where the rate of supply of the product is constant, either by virtue of the long life of the parent or because the quantity of the parent is itself maintained by an earlier disintegration. If otherwise, and the product is longer lived than the parent, then, of course, after the whole of the parent has disintegrated the quantity of the product diminishes with its own proper rate of change. Another consideration then comes in. If the period of the parent is very short, the radioactivity of the longer lived product may not be sufficiently great to be detectable. Thus an intense activity due to radium C, which decays rapidly, gives rise to a very feeble but enduring activity due to radium D and its products, the period of radium D being a quarter of a million times greater than that of its parent. The effect of the α-activity of radium A deposited from the radium emanation on the walls of the containing vessel during a few minutes is comparable with that given by the emanation itself, whereas the emanation of thorium has to be maintained in a continuous stream for many hours before an effect of the thorium active deposit at all comparable with that of the emanation is produced on the walls of the vessel. This illustration is complicated by there being a rayless change in the latter case, but the main reason is that in the first case the period of the parent is about 2000 times greater than that of the product, whereas in the second example the period of the product is about 700 times that of the parent. All of these illustrations are merely several examples of the fact already alluded to that the quantity of radioactive substance divided by its period of average life fixes the number of atoms disintegrating per unit of time, and hence the degree of radioactivity. Since in the majority of cases the quantity of the matter concerned in the phenomenon is below the quantity detectable by any test except radioactivity, the quantity itself is of no importance. What is of importance is the radioactivity or quantity divided by the period. No product of uranium X has yet been detected, although there is every indirect reason to believe that the α-ray giving ionium is produced. It can be calculated from certain experiments in which a growth of the α-rays of ionium from uranium X, separated from 50 kilos. of uranium nitrate, was looked for and not found, that if ionium is the direct product of uranium X, its period must be at least 30,000 years. The same holds with regard to the ultimate inactive products at the end of each series. If these are not wholly stable elements, the periods must be extremely long, or they must disintegrate raylessly. Otherwise they would be detectable.

CLASSIFICATION AND NOMENCLATURE OF THE RADIO-ELEMENTS—ANALOGIES BETWEEN THE THREE DISINTEGRATION SERIES

THE thirty or more radio-elements known may be classified in various ways. They are all almost certainly derived from one of two primary parent elements—uranium and thorium. One of the commonest and most useful methods is to arrange them in two disintegration series, starting with these two elements in order of their production. This is the order in which the substances are taken in the sequel. The Table at the end of the book comprises all the radio-elements known, arranged in this way. It includes the period of average life of each member and the nature of the radiation, if any, expelled when it disintegrates and changes into the next member of the series.

In the uranium series, in addition to the main series, comprising ionium, radium, and polonium, there is a subsidiary series consisting of actinium and its products. The manner in which actinium comes to be produced is still not completely elucidated. Only a very small proportion of the uranium atoms disintegrating appear to go to produce actinium, by far the greater proportion probably going by the main series ultimately into polonium and probably lead. At some point in the series, which has not yet been determined, the atom disintegrating most probably has two modes of disintegration with different periods and different products, some of the atoms giving one and the others the other product. It is practically certain that the product of polonium is lead, although this has still to be established directly. The ultimate products of the thorium and actinium series, however, remain quite unknown. In the diagram the numbers refer to the atomic weights of the successive products. When these are in brackets it indicates that the atomic weight has been calculated from that of the parent and the number of α-rays expelled in the formation of the product.

The nomenclature adopted has gradually been evolved and is temporarily accepted until a complete scientific system can be formulated; it is still being frequently altered as new discoveries are made. The names chosen for the new radio-elements which were discovered before their position in the scheme was known have been usually retained, and sometimes new products in one series have been named by analogy to similar products in one of the others. In the case of the end products of disintegration of radium, which were worked out comparatively late, the system has been adopted of calling them in sequence radium A, radium B, radium C, &c. Sometimes, as in the case of radium C, it has been found that what has been considered a single change consists of two. In some cases the two separate products are not individually known, their separate existence being only indirectly inferred. This is provided for without disturbance to the rest of the names in the series by calling the first body, for example, radium C_1 and the second radium C_2. Mesothorium 1 and mesothorium 2 constitute another such instance.

Another possible method of classification of the radio-elements is based upon their periods of average life or rates of change. The parent elements, uranium and thorium, have the longest periods and change the most slowly, so that their chemistry does not differ at all from that of the inactive elements. If their products are then considered without reference to the series to which they belong, in descending order of their periods of life, we pass gradually through a series of substances in which the chemical methods of investigation become more and more different from those usually employed and more and more peculiar to the study of radioactivity. The practical methods of preparation and purification of the radio-elements are entirely controlled by the period of average life of the substance.

In a mineral the various members of each disintegration series exist in quantities directly proportional to the period of average life, so that a radio-element like, for example, polonium with a period of life of less than a year must be present in the mineral in far smaller quantity than radium which has a period some thousands of times longer. With the short-lived substances, with periods of only a few hours or minutes, the actual quantities dealt with are always quite infinitesimal and quite incapable of being detected except by their radioactivity. A complete knowledge of the chemical behaviour of a radio-element in presence of any possible mixture of other elements, such as is involved, for example, in its extraction from the original mineral, is only required and is only possible for those radio-elements of period of the order of a year or more. This class comprises some nine radio-elements, which, arranged in descending order of period, are Thorium, Uranium, Actinium, Ionium, Radium, Radio-Lead (Radium D), Mesothorium 1, Radio-Thorium, and Polonium. These are usually derived directly from the mineral, and indeed for most this is the only way of obtaining them. Radio-elements of shorter life than these cannot well be separated from the original minerals, if only because during the progress of the necessary chemical operations they tend to change as fast as they are separated from their parent substance. They are invariably obtained by separating the parent in as pure a state as possible, in the radioactive sense—that is, as free from other radioactive matter as possible, and allowing it to accumulate its short-lived products. For the separation of the latter it suffices to use means of separation capable of removing them from their parent on the one hand and from their own products on the other. Such operations are often relatively simple and rapid, and are often effected by adding a sensible quantity of an inactive substance which has a chemical resemblance to the radio-element required, which is only present in infinitesimal amount, and so using the former as a vehicle in the separation of the latter. Thus uranium X is prepared from a purified uranium salt which has been kept a few months since preparation, and has regenerated its full equilibrium quantity of uranium X. Thorium X is prepared similarly from pure thorium salts (which always contain radio-thorium, the immediate parent, in considerable quantity), and so on. In the case of the two last-mentioned members of the first class, radio-thorium and polonium, the first cannot be separated chemically from thorium, and so must be obtained from its parent mesothorium (which is separable from thorium) after it has been prepared for some time, while a valuable alternative source of polonium, instead of obtaining it direct from the mineral, is an old preparation of radio-lead in which polonium has been regenerated with lapse of time. Radio-lead, again, may be obtained from an old radium preparation which has been suitably preserved.

Finally, owing to the existence of the gaseous emanations in each series, which are very readily removed from all non-volatile matter, and can, if desired, be condensed out of gases at the temperature of liquid air, they and their subsequent products can be readily obtained practically pure in unweighable and invisible quantities, but of extraordinarily intense radioactivity. The chemistry of such emanation products, or "active deposits" as they are called, involves only the separation of the successive products from one another. However short-lived, they can be separated in the pure state by physical methods alone, which in other cases would be difficult or quite impossible.

Active Deposits.—There are three "active deposits" or non-volatile products of the three emanations. The latter can readily be removed from the preparations which produce them mixed with any required gas, either by the ordinary methods of gas manipulation, as in the case of the comparatively long-lived radium emanation, or by passing continuous streams of gas through the preparation, preferably in solution, in the case of the shorter-lived emanations. The products of the emanations are, therefore, produced in gases away from all solid or liquid materials. Their collection and concentration on a metal surface or point is uniformly and simply effected by charging the surface in question negatively with respect to the surface of the containing vessel. It is probable that all products of disintegration, so formed and free to move in a gaseous atmosphere, possess initially, or acquire instantly, a positive charge. Even when the radiant particle expelled in the same disintegration is an α-particle, and is therefore positively charged, the charge of the product, or residue of the disintegrating atom, is always positive. The explanation of this apparent anomaly is, probably, that in neither case is the positive charge initially present after disintegration. The collision both of the radiant particle and of the recoiling residue of the atom after disintegration with the gas molecules in their path would cause them to acquire positive charges, even if initially uncharged.

If it is desired to collect the active deposit from the thorium or actinium emanations, the latter may be passed by a current of air into a metal cylinder provided with a central electrode of wire along the axis held in place and insulated from the cylinder by rubber corks. The central electrode is attached to the negative pole and the outer cylinder to the positive pole of any source of potential, such, for example, as the electric light mains, a lamp being inserted in the circuit in case of an accidental short circuit. In most cases 50 volts or even less is ample for the purpose. To collect the radium active deposit, the emanation may be stored in a silvered glass flask, the internal coating of which is made positive, and a wire or knitting needle, sheathed in a glass tube to expose only the length required to be made active, and held in the flask by a rubber cork, is made negative.

The potential gradient, or volts per centimetre of distance, between the electrodes must be increased if the intensity of the radioactivity is very great. As a general rule it may be stated that the potential gradient required to concentrate the whole of the active deposit on the negative electrode is the same as would be necessary to produce "saturation" of the ionisation current through the gas under the same circumstances, that is, to collect all the ions in the gas at the electrodes. Higher potential gradients than 50 volts per cm. are rarely necessary except in the case of intensely radioactive preparations, or where the concentration is to be effected on a very small area of surface.

Under these circumstances the active deposit is formed entirely on the negatively charged surface, which may be removed at any time from the emanation and studied by itself. Often the active deposit is then dissolved off in hydrochloric or nitric acid, or volatilised in whole or part by exposure to a high temperature, so that in such cases the negative electrode is preferably made of platinum.

In the systematic account which follows, the radio-elements will be considered in their order in the disintegration series, commencing with that of uranium, next taking that of thorium, and lastly that of actinium. In each series the "active deposit" will be considered as a group for the sake of brevity.

The new nomenclature (October 1911) recommended by Rutherford[19] in consequence of the discovery of important new members has been followed, and it must be pointed out that this involves serious changes. Thus what has previously been termed thorium A and actinium A are now called thorium B and actinium B, and so on through each series. This is at first sight very confusing, but it leads to some important advantages, of great assistance to the memory in giving the right name to the right substance, so that its temporary disadvantages are more than counterbalanced. These advantages will be referred to when the analogies between the series are discussed (p. 27).

The Relation between Radio-Chemistry and Chemistry.— Uppermost in the mind of any one approaching the subject from the point of view of chemistry is the question as to how far such knowledge as can be acquired of the chemical properties of these short-lived radio-elements can be accepted literally, and used as a basis of classification of the radio-elements in the same way as the common elements have been classified in the Periodic Law, for example. It may be urged that although an infinitesimal quantity of ionium, to take one case, behaves, when mixed with a relatively enormous excess of thorium, like the element thorium itself in all respects, the chemical nature of ionium, if it were possible to prepare it free from thorium, might be quite different. Every worker at the present time no doubt has formed some more or less tentative opinion on this important point, and such opinions, if set forth, would no doubt be very different according to the personal experiences of the worker. The one expressed in this section must be taken rather as an individual opinion than as a settled judgment. It may be remarked that the common present-day method of separating a radio-element from a mixture by adding deliberately its nearest chemical analogue and separating this, has rather tended to close the avenue of knowledge as to the real chemistry of the radio-elements, and for this reason the work of Auer von Welsbach[20] under the Austrian Government on the pitchblende residues, in which this method was purposely not employed, is of considerable interest. The amounts of thorium, for example, which he separated from the residues of many tons of pitchblende were so small, that there is little doubt that they must have contained a considerable percentage, if not actually a predominance, of its more radioactive twin-brother ionium.

The vague idea that infinitesimal amounts of radioactive matter may possess a chameleon-like nature, reflecting the properties of the substances with which they are mixed rather than their own, is not justified. The properties of the radio-elements, and with certain reservations to be referred to, their behaviour under any defined set of conditions, is at least as definite as those of the common elements. The statement that ionium cannot be separated from thorium, and many similar ones to be detailed in the sequel, is tantamount to saying that these radio-elements *can* be separated completely from all the other elements differing in chemical reactions from the one specified. Admittedly many have not, and perhaps never will be separated from the element they most resemble. But it is inconceivable that a radio-element when pure might possess chemical properties, allied most closely to those of, say, lanthanum, and be capable of being separated from lanthanum but not from, say, thorium out of a mixture containing both of these elements.

A little consideration will serve to show that those radio-elements with chemical and physical nature indistinguishable from a common element always present with them in the natural minerals will behave definitely, whereas those others which are not allied to any known common element in properties will probably behave the more indefinitely the shorter their period of life. The phenomenon of "adsorption" at the surfaces of solids is probably a general property of all surfaces, which varies enormously both with the nature of the surface and the nature of the substance adsorbed. In the analytical reactions of the common elements it is only remarked when the adsorbing surface can take up weighable quantities of the adsorbed materials; quantities, that is, which correspond to more than 10^{18} molecules. A given quantity of an adsorbent might be able, for example, to adsorb a million molecules of most substances, and more than 10^{18} molecules of one particular substance or molecular species. It would then be regarded as an adsorbent of the one molecule alone. In a solution containing a radio-element the adsorbent might adsorb the radioactive substance, if present in quantity below a million molecules, completely, provided only that no other similar molecular species was present. But if the radio-element were physically and chemically indistinguishable from some common element, which was present in far less than weighable quantity, the adsorption of the radio-element might practically be prevented. An example of this will be found under *Uranium X*, the adsorption of which by charcoal is prevented by minute traces of thorium. This suggests a general method of determining whether a short-lived radio-element is indistinguishable from any one of the known elements, for if so, its adsorption should be prevented by the presence of that element.[21]

One of the earliest examples of the extremely sharp and definite behaviour of almost inconceivably small amounts of the radio-elements was furnished by the condensation of the emanations at low temperature. Since then many equally definite examples of differences in volatility of successive products in a series have come to light, and indeed this affords one of the general methods of separation employed. Often, no doubt, there are causes which contribute to make the behaviour of the radioactive substance more definite than could be foreseen, and one of these is undoubtedly the presence of other substances with similar properties. Thus in a solution containing barium and radium, the radium may be precipitated quantitatively by sulphuric acid although present in quantity so minute that it is inconceivable that the radium sulphate would not be completely soluble if the barium were not present. It is however significant that the radium in such a system will redissolve again in the course of time if alternately heated and cooled and agitated.[22] So also in many other cases. Often the behaviour is, with minute quantities, unexpectedly definite, but indications of the chemical and physical nature drawn from the study of these minute quantities are not usually found to be at fault, when the behaviour of larger quantities of the material can be subjected to examination.

Analogies between the Disintegration Series.—The disintegration series as they are now known and represented in Fig. 1 offer a glimpse at once as arresting and as elusive as that afforded by the Periodic Table into the unsolved problem of the nature of matter. There is a general analogy between all three series.[23] That between actinium and thorium is almost perfect, except that two products are known between thorium and radiothorium of which no representatives occur in the actinium series. The uranium series is distinguished by three products more at the end unrepresented in either of the other series.

These analogies have become more obvious in consequence of certain discoveries already alluded to.[19] It has been established that the thorium and actinium emanations both disintegrate with the production of excessively short-lived non-volatile products which give α-rays, but which before have not been separately distinguished from the volatile parent. In consequence, the first member of the "active deposits" in each of the three series is now a short-lived α-ray giving product radium A, thorium A, actinium A. The second or B member is a rayless, or more strictly, a (β)-ray giving product, having the longest period of the active deposit group in each case. After this the analogies, though still striking, become less perfect, and it is scarcely profitable, in view of probable new discoveries in this field, to discuss them at length. The next, or C member, in each case gives α-rays, and is somewhat shorter-lived than the preceding member. In all cases it is probably not a single product but consists of two products, indicated by C_1 and C_2. It is doubtful whether these are successive, or simultaneous products. Thorium is the simplest case. The complexity is indicated by two different sets of α-particles being emitted, the range of the C_2 set, 8.6 cm., being the longest known. The product is therefore probably of unmeasurably short period. In actinium indications of the same sort exist, but they are less definite, as the ranges of the particles are nearly alike. In the case of radium only one set of α-rays are emitted, and the range is so high, 6.6 cm., that probably these do not come directly from C_1 but from another very short-lived product. The name, radium C_2, however, has unfortunately been given to a short-lived product giving only β- and γ-rays, and possibly a branch product of C_1. Radium

C_1 also, curiously, seems to give the same kind of β- and γ-rays as C_2. In the case of thorium and actinium the typical penetrating β- and γ-rays are given by a subsequent D product of fairly short period. If in the radium series a separate β- and γ-ray product is ultimately discovered it should be called radium D, and the subsequent products correspondingly renamed. However, radium D is the name given to the long-lived product, giving only feeble (β)-radiation, which is obviously analogous to the stable ultimate E products of the thorium and actinium series, which have not yet been identified. Since radium D is chemically indistinguishable from lead, by analogy the other two products should be stable elements in this part of the Periodic Table.

Disregarding then the minor differences, it may be stated that the A members give α-rays, and are very, sometimes excessively, short-lived, the B members are the longest lived of the active deposit group, and give only unimportant (β)-radiation, the C members give α-rays and are complex, the first representative being rather shorter-lived than the B members. The D members (except radium) are short-lived, and give the typical β- and γ-rays. The atoms then re-emerge into the stable or comparatively stable region around lead and bismuth in the Periodic Table. This new nomenclature assists the memory because it makes the names of the analogous members of the "active deposits" correspond in the three cases. Obviously, however, it would be more perfect if the nomenclature had anticipated future discovery by changing the name of radium D to radium E, and of its products correspondingly (October 1911).

After radium D comes a short-lived β-ray product radium E, then the penultimate α-ray giving radium F or polonium, the position of which in the Periodic Table can be fixed from its estimated atomic weight and chemical nature, as occupying the vacant space in the sulphur-selenium-tellurium group of elements, next in atomic weight to bismuth. The ultimate product, radium G, is almost certainly lead. In the thorium and actinium series these members are not represented, the series ending apparently with the E members.

Working now backwards through the series, we have first the three emanations, all of which belong to the zero group of the Periodic Table, and are chemically inert like argon.

Next come radium, thorium X, and actinium X, three chemically identical and non-separable elements of the alkaline-earth group. Next come ionium, radio-thorium, and radio-actinium, the latter of which is somewhat indefinite in chemical nature, but the first two of which are identical chemically with thorium of the carbon or tetravalent group and not separable from it. Now another set of suggestive differences in the series appears. Actinium and mesothorium 1, the only two really rayless members known, find no analogue in the uranium series, and probably a long-lived rayless product, somewhere intermediate between ionium and uranium, remains to be discovered. Uranium X seems to correspond well with mesothorium 2, both being fairly short-lived and giving penetrating β-rays.

Already, therefore, the disintegration series affords a most remarkable picture of the actual process of the production of the elements from one another, of which the Periodic Law is, as it were, the consequence. Just as from an instantaneous photograph of a waterfall the movement of the apparently motionless water can be inferred, so from the Periodic Law the continuous transformation of the apparently unchangeable elements has been suspected.

Radioactivity has, as it were, cinematographed these transformations, with the result to-day, which none ten years ago could have dared to imagine possible, that in three separate instances we are tracing the successive transit of matter from group to group of the Periodic Table. There are certain points in particular that appear of great significance. The loss of a helium atom or α-particle appears to cause the change of the element, not into the next family but into the next but one. All the products known well enough to classify are of even valency, and this in spite of the fact that the atomic weight of thorium is some six units (or one and a half times the atomic weight of helium) less than that of uranium, and more than that of radium. The families of odd-valency are nowhere represented. Thus we have in each series a well-marked sequence from the tetravalent family (radio-thorium, ionium) into the divalent family (radium, &c.), and into the non-valent family (emanations). Again, the product derived from polonium (group VI.) by the loss of an α-particle is probably not bismuth but lead, in each case the step being from the family of even valency into the next, the family of odd valency being missed. But this is not all. The progress is certainly not so straightforward as this. In several cases the matter appears to *alternate* in its passage, passing through the same family not once but *twice*. The product from thorium (group IV.) is mesothorium (group II.). The product of the latter is radio-thorium (group IV.), which, in turn, produces thorium X (group II.). Again radium D is chemically non-separable from lead (group IV.), its product is polonium (group VI.), while the product of the latter is almost certainly lead (group IV.).

In this connection it should be noted that the helium produced by these changes, which in its ordinary state appears non-valent like argon, carries, in the form of the radiant α-particle, two atomic charges of positive electricity, and is therefore electro-chemically divalent. The passage from group II. (radium) to the zero group (emanation) is direct, whereas the passage from the zero group, the beginning of one period, back to the end of the last period occurs through the long chain of "active deposit" products. In the Periodic Table the transition appears abrupt, and no indication is afforded of intermediate connecting links. The active deposits thus represent a new region in the constitution of matter, of the existence of which there has so far been no evidence. This region bridges the two ends of the Table. The atom having suffered successive reductions of its valency to zero, passes to the electro-negative end of the preceding period through a well-defined sequence of short-lived forms. According to von Lerch's rule, the process is accompanied by a regular increase in the electro-negative character, the successive products being each electro-chemically "nobler" than the last. Obviously the study of the disintegration series is affording fresh light on the Periodic Law, and is showing that new kinds of atoms, nowhere represented by that law, can and do exist, though often they are excessively unstable.

URANIUM (U)

Period of average life—About 8,000,000,000 years.
Radiation—α-rays. (Two α-particles per atom of uranium.)
Range of α-rays—2.58 cm. of air.
Disintegration product—Uranium X.

Occurrence.—The most important deposits of uranium, radium, &c., occur as pitchblende or uraninite (chiefly U_3O_8) at St. Joachimsthal, Bohemia, which, however, is supposed to have been worked out; at Johann Georgenstadt, Saxony; in Cornwall, England; in North Carolina and Connecticut, U.S.A.; and in German East Africa. Another important ore is carnotite (uranium potassium vanadate), found chiefly in the U.S.A. Almost all the thorium minerals (*q.v.*) contain a certain quantity of uranium and radium. In thorianite, which contains about 60 % ThO_2, the quantity of uranium is usually either 10 % or 20 %.

Properties.—Like iron, uranium forms two series of salts corresponding with the oxides UO_2 and UO_3. The common black oxide U_3O_8, formed by heating other oxides in air or oxygen, is regarded as a compound of these, $UO_2 + 2UO_3$. The higher oxide is acidic, like chromic oxide, CrO_3, and forms salts with alkalies, known as the uranates, having usually analogous composition to that of the dichromates, as, for example, $Na_2U_2O_7$ and $K_2U_2O_7$. Uranium is, however, less acidic than molybdenum and tungsten, the other members of the family. Salts of the lower oxide UO_2, as, for example, UCl_4, are not usually met with in ordinary work, as, like ferrous salts, they are powerful reducing agents, and are readily oxidised even by air. In absence of iron, uranium may be conveniently estimated by reducing it in sulphuric acid solution by zinc or magnesium, and tritrating the solution with permanganate. The uranous salts are usually green. The commonest compounds of uranium are the uranyl compounds, which are usually yellow. These contain the divalent radical (UO_2), and are derived from the higher oxide UO_3, just as sulphuryl chloride SO_2Cl_2 may be regarded as derived from SO_3. Thus in acting on the higher oxide UO_3 with hydrochloric acid, the product is uranyl chloride UO_2Cl_2. Uranyl nitrate, $UO_2(NO_3)_2 + 6H_2O$, is the commonest salt, and is distinguished by its magnificent crystals, showing a yellow-green fluorescence, and by the readiness with which it crystallises from concentrated aqueous solution. This is the method by means of which uranium salts are usually purified on a large scale. The crystals melt in their water of crystallisation at about 50°, and lose their water and decompose, with evolution of nitrous fumes and formation of UO_3, when heated gently on the sand bath. The salt, both hydrated and anhydrous, dissolves readily in common ether, and this property is frequently made use of in analysis and small scale preparations for separating uranium from minerals. On a large scale the distillation of the ether from the solutions of uranyl nitrate is apt to result in violent and even dangerously explosive oxidation of the ether unless large quantities of water are added before distillation. In this reaction an amorphous yellow insoluble compound of uranium having the approximate composition UCH_4O_6 is formed. It decomposes readily at 220°–300°, with loss of CO_2 and H_2O, forming a basic uranous carbonate. It is soluble in ammonium carbonate solution, and has not been previously described.

The property of most value, however, in the separation and analysis of uranium, is the solubility of the uranyl salts of all the common weak acids, such as carbonate, phosphate, &c., in alkaline carbonates. Ammonium carbonate is usually employed, as this reagent on heating dissociates into ammonia and carbon dioxide, which completely escape from the solution, so that the uranium is reprecipitated completely by simple evaporation. But sodium and potassium carbonates are equally effective, and are sometimes preferable.

Separation and Estimation.—In the technical assay of uranium in pitchblende, what is known as Patera's method is practised at the State mines of St. Joachimsthal. The pitchblende is dissolved in nitric acid, poured into moderate excess of sodium carbonate and filtered, preferably after standing for some hours to ensure complete separation of the iron. The operation may be repeated on the precipitate, if necessary, but under proper conditions the whole of the uranium is contained in the solution practically free from other constituents of the mineral, and may be precipitated as sodium uranate by acidifying, boiling off the carbon dioxide completely, and adding sodium hydrate. The precipitate is filtered and ignited without washing, washed free from alkali, and weighed as sodium uranate.

This pouring of a solution of uranium containing a complex mixture of other constituents into excess of an alkaline carbonate is always a valuable first step in the separation of the uranium, and when the latter element is the only one being separated or estimated, it affords a ready means of obtaining the whole of the uranium in the filtrate in a state of approximate purity. For analytical estimations it is almost always an advantage to add some ammonium sulphide to complete the precipitation of the iron. Any suitable method can then be employed for the separation and estimation of the uranium in the filtrate. One of the most commonly employed is, after acidification and removal of the carbon dioxide, to add microcosmic salt (sodium ammonium hydrogen phosphate), to neutralise as nearly as possible with ammonia without causing a permanent precipitate to form, and then to add a considerable excess of sodium thiosulphate. Boiling for ten minutes brings down the whole of the uranium as phosphate with a bulky sulphur precipitate which settles and filters excellently. After washing, the precipitate is ignited at a dull red-heat and weighed as the "green phosphate," a body which has been found empirically to correspond with 68.55 % of uranium, and which is far less hygroscopic than the pyrophosphate. When pure it dissolves completely in concentrated nitric acid, but when impure the solution is more difficult. Evaporation of this solution and ignition leaves the yellow pyrophosphate $(UO_2)_2P_2O_7$, which contains 66.81 % of uranium. The conversion of the "green phosphate" in the crucible into the yellow pyrophosphate affords a check on the estimation, but the latter body is excessively hygroscopic and difficult to weigh. (For further details, Brearley's *Analytical Chemistry of Uranium* should be consulted.)

Uranium may also be precipitated as ammonium uranate, and ignited and weighed as U_3O_8. The only difficulty is the washing of the precipitate, which must be done with a 4 % ammonium nitrate solution, as water causes the precipitate to become colloidal and run through the filter. Another method is to precipitate uranium with ammonium hydrate and sulphide, as uranous oxide UO_2, and to weigh as such after ignition in hydrogen, or as U_3O_8 after ignition in oxygen. Autunites and similar phosphates containing calcium are best estimated as phosphate, acetic acid being present to prevent the precipitation of the calcium.

An excellent colorimetric method of estimating uranium, applicable to very dilute solutions, is to add to the solution and to a similar volume of water in small evaporating basins the same amounts of potassium carbonate solution and half a gram or so of sodium peroxide. The intense yellow colour due to uranium is visible in extremely dilute solutions, and is more easily matched when faint. The colour in the basin containing the uranium is then matched with that of the blank by running into the latter standard uranium solution. Iron interferes badly with the tint. It is best removed by treating the uranium solution with the ammonium hydrate, carbonate, sulphide mixture as described. A tenth of a milligram of uranium can readily be detected, and, with care, estimated in this way. The method is of value in the rapid examination of precipitates, &c., obtained during the separation of uranium, to test whether all the uranium has been removed.

Treatment of Radioactive Minerals.—The radioactive constituent of uranium minerals of greatest technical value is, of course, radium, whereas the valuable constituent in the case of thorium minerals is mesothorium, which has identical chemical properties. Hence they are always separated together, if present, and the same treatment is applicable to both. The older methods were largely developed in consequence of the existing practice in the treatment of uranium minerals. These were fused with sodium sulphate, and the uranium lixiviated out with water and dilute sulphuric acid, the insoluble residues containing the radium and radiolead as sulphate, and part of the ionium, polonium, and actinium. Probably these residues are now no longer obtainable. The separation of the radioactive constituents has become at least as important as the separation of the uranium, and starting with fresh mineral *ab initio*, far simpler methods could be devised according to the character of the mineral. Many of the minerals, for example, carnotite, autunite, and thorianite, are soluble in dilute acids, and from such solutions, by pouring into sodium carbonate, the uranium can be at once recovered from the filtrate, whilst the radium and other active constituents are precipitated as carbonates in a form very easily redissolved and further concentrated. In the case of pitchblende, which is soluble in nitric acid, the existence of sulphides causes sulphates to be produced, and part of the radioactive materials are precipitated as sulphates. But such precipitation can often be prevented by use of concentrated acid until the solution has been separated from the bulk of the insoluble material, and the amount of material to be worked up for radium could no doubt be enormously reduced if desired. But sooner or later the separation of the radium and mesothorium, if present, which constitute the two most important of the radioactive constituents, is effected through the insoluble sulphates and their separation with barium sulphate. This method will be referred to under *Radium*, but for full details of the factory processes the originals must be consulted.[24] These processes appear unnecessarily prolonged and tedious, and are no doubt largely obsolete. When the active elements have been precipitated with the insoluble sulphates a considerable concentration can often be effected before solution of the active material by treatment with 25 % sodium acetate solution, which dissolves the lead sulphate and part of the calcium sulphate.

Radioactivity.—All uranium salts, and uranium itself, in ordinary circumstances give, in addition to the α-radiation, an important and fairly penetrating β-radiation, due to uranium X. The latter is easily separated in most of the processes of purification in vogue, so that freshly prepared uranium salts give at first only α-rays. The β-rays are gradually regenerated. After about three weeks from purification they are nearly half the ordinary equilibrium strength, equilibrium being practically reached after from six months to a year.

Standards of α-Ray Activity.—On account of the constancy of the α-radiation of uranium, which, till now, has not been diminished or affected by any chemical process, the most convenient standards of radioactivity to employ in the standardisation and calibration of electroscopes and other measuring instruments employed in radioactive work, are made of thin films of black oxide of uranium U_3O_8. The commercially pure uranyl nitrate may be further purified by one of the methods previously alluded to, either by crystallisation, solution in ether, or pouring into ammonium carbonate. It is then cautiously ignited in a platinum crucible. The ignition is continued on the blast in a current of oxygen, which is maintained after the heating is stopped. The pure black oxide is ground up in an agate mortar with freshly distilled chloroform to a thick paste. By means of a clean camel-hair brush, shallow circular copper trays of definite area are coated with the smallest quantity of the paste necessary completely to cover the surface, and the chloroform is then allowed to evaporate spontaneously. The α-radiation from the trays is proportional to their area, and, so long as the surface is completely covered, is quite independent of the amount of material used to cover them. Provided no more of the uranium oxide is used than is necessary to cover the surface, the β-radiation is practically negligible. For very accurate work it is advisable to use a preparation of uranium oxide, which has been prepared for some months, of which the β-radiation is constant. Such standards are, of course, used bare, but are kept covered to protect them from dust when not in use. They are easily prepared, and the film adheres well to the surface. By preparing similar films of minerals or other materials and comparing their discharging action on an electroscope with that produced by the standard uranium film, the α-activity of the substance in terms of that of U_3O_8 may at once be accurately determined by means of a simple electroscope.[25] The total α-activity of a uranium mineral is 4.64 times that of the uranium it contains if there is no thorium present. Hence by dividing the activity of the mineral in terms of that of U_3O_8 by this factor, a rough evaluation of the proportion of the contained U_3O_8 may be made.

Uranium and uranium X form for practical purposes a distinct and complete disintegration series in themselves, which is connected with the long ionium-, radium-, radiolead-, polonium series on the one hand and the actinium series on the other only by indirect theoretical reasoning. For in spite of the most exhaustive and delicate experiments, no direct production of ionium, radium, or actinium has so far been observed from uranium. Neither from the uranium X, separated from 50 kilograms of pure uranyl nitrate, has there been observed any production of a radioactive product, though, theoretically, ionium should be formed from uranium X either directly or through some unknown intermediate product or products.[26] For all practical purposes the decay of the activity of uranium X is as complete as that of polonium, the end member of the radium series, or of thorium C, the last member of the active deposit and of the series of thorium. With this may be contrasted the decay of radium C, the last member of the active deposit of radium, the decay of which is only apparently complete. Really the activity falls to a very small value and then steadily increases again for many years, due to the formation of later products which are radioactive. The extreme simplicity of the uranium series proper, which contains only two radiating substances, one giving all the α-rays and the other all the β-rays, the comparatively short period of the β-ray producing product, and the absence of all slowly changing or gaseous products of disintegration, make the compounds of uranium especially adapted for use as standards of radioactivity and for illustrating the main principles of radioactive change in their least complex aspect.

The fact that uranium expels two α-particles per atom disintegrating, whilst all the other members of the uranium-radium series,[27] giving α-rays on disintegration, expel only one, makes it possible that uranium is not a single element, but a mixture of two, chemically non-separable, differing in atomic weight by four units, and both expelling α-rays. Such a mixture would act as a single element expelling two α-rays per atom. However, in the thorium and also the actinium series the rule of one α-particle per atom disintegrating does not appear to be followed.[28] This makes it impossible at present to estimate the atomic weights of the members of the thorium series and of its ultimate product. Analogies between the series suggest that there should be also a rayless product between uranium and ionium corresponding to mesothorium and actinium, and, indeed, the delay in the appearance of radium in pure uranium preparations is so prolonged that at present it suggests the existence of long-lived intermediate bodies other than ionium.

URANIUM X

Estimated atomic weight—230.5.
Period of average life—35.5 days.
Period of half-change—24.6 days.
Radiation—β, (β)- and γ-rays.
Penetration power of rays—Half absorbed, β- by 0.434 and (β)- by 0.0102 mm. of aluminium, γ- by 0.95 cm. of lead or 1.68 cm. of copper after the first 1 cm. of thickness.
Nearest chemical analogue—Thorium.
Parent—Uranium.
Disintegration product—Unknown.

This constituent is responsible for the penetrating rays of ordinary pure uranium compounds, and may be easily separated by many of the methods of purification described under *uranium*. Thus, on crystallising uranyl nitrate by allowing a hot solution, of sp. gr. 2.05, to cool, about two-thirds of the salt crystallises, and the mother liquor contains about six-sevenths of the uranium X present.[29] On shaking crystals of the same salt with ether, two layers are formed, of which the lower aqueous layer retains the whole of the uranium X. Other solvents, such as acetone, various alcohols, ethyl and methyl acetate, have been recommended, but it is doubtful if they offer any advantage. In all such cases the solution in the organic solvent may be shaken with freshly precipitated ferric hydroxide and filtered, which makes the separation easier in unpractised hands. On pouring a uranium solution into excess of a solution of carbonate of sodium or ammonium, the iron, alumina, and other impurities are precipitated, and these retain the whole of the uranium X. If the uranium is too pure it is necessary to add a little ferric salt to serve as a nucleus for the uranium X. It may be separated from iron by dissolving the precipitate in concentrated hydrochloric acid and extracting the solution with ether saturated with hydrogen chloride, when the uranium X remains in the aqueous layer.[30] In the ammonium carbonate process the uranium X precipitate, if too great, may be redissolved in acid, and poured into very concentrated ammonium carbonate, and the solution reprecipitated fractionally by boiling. According to the nature of the other substances present, the uranium X comes down sharply and definitely with one of the fractions, usually in the middle of the series.[31]

Uranium is readily adsorbed by the action of barium sulphate formed in the solution or by carbon. Soot, obtained freshly by burning napthalene, has been recommended. These adsorptions seem to be entirely prevented by the presence of thorium, which is difficult to account for unless uranium X and thorium are, chemically and physically, practically identical.[32] No full investigation of this point has been published, but the addition of thorium to the uranium solution and its precipitation in acid solution with hydrofluoric acid has been recommended as a method of separation of uranium X from uranium.[33]

If the uranium X is required as free as possible from all other substances a carefully purified rare-earth, such as lanthanum, may be added instead of thorium. After separation the solution of the rare-earth is precipitated with sodium thiosulphate. Sufficient thorium is usually present to ensure that the uranium X is precipitated with the sulphur formed, so that after ignition only a very minute residue is left.[34]

The radioactivity of uranium X is characterised by its homogeneous penetrating β-radiation, which is only slightly less penetrating than that of radium C. The γ-radiation is extraordinarily feeble, as compared, for example, with preparations of radium, mesothorium, or radio-thorium of equal β-activity, and is less penetrating than the γ-rays of these substances, especially for thicknesses less than 1 cm. of lead or its equivalent.[29] The (β)-rays were at one time mistaken for an α-radiation. From a bare thin film they produce an ionisation rather greater than that due to the β-rays, but they are practically completely absorbed by 0.05 mm. of aluminium.

The disintegration product of uranium X, in spite of special investigations with enormous quantities of uranium, remains unknown.[35] Theoretically it should be, or should produce ionium, but this has not yet been detected. Uranium X has not been volatilised.

Uranium Y.—At the time of writing, a substance, termed uranium Y, has been described, characterised by a greater proportion of (β)-radiation than uranium X and a very feeble α-radiation.[84] Its period is 1.5 days, and although the evidence is very far from complete, it is regarded as being probably a product simultaneously formed with uranium X in relatively exceedingly minute quantity. Its properties are very similar to those of uranium X. It is separated with the uranium X by the ferric hydrate, but not by the barium sulphate method.

IONIUM (Io)

Estimated atomic weight—230.5.
Period of average life—Probably between 50,000 and 1,000,000 years.
Radiation—α-rays.
Range of α-rays—2.84 cm. of air.
Nearest chemical analogue—Thorium.
Parent—Uranium X (?).
Disintegration product—Radium.

The existence of this substance, an intermediate product of long-life between uranium X and radium, was foreshadowed by the failure to detect a growth of radium in uranium preparations purified from radium. It was discovered and isolated from pitchblende and other uranium minerals by Boltwood.[86] In the preliminary treatment of the mineral it is separated with the actinium, and indeed the first actinium preparations separated owed their radioactivity almost certainly in part to admixture with ionium. It is, however, easily separated from actinium.[87] Its chemistry can be fully and accurately described in a single sentence. It resembles thorium in its whole chemical nature, as far as it is known, with absolute completeness, so that not only is no separation of the two elements possible by any known method, but no concentration of the one constituent with reference to the other has been accomplished. Such an alteration of concentration could be readily detected, if it occurred, even with the infinitesimal amounts of ionium used, because of the intense α-activity of the latter. Thorium and ionium form a pair of non-separable elements. Radiothorium, certainly, and, according to some authorities, uranium X, are also non-separable from thorium. This apparently complete chemical identity of different elements of, in some cases, different atomic weight, is one of the most important recent results of the study of radioactivity, and many other examples will occur. It must be remembered that the chemistry of thorium has been closely and exhaustively studied on account of its resemblance to and association with many other rare-earths of no technical value, and in consequence a large number of characteristic and effective methods for the purification of thorium from all other elements are known (see *Thorium*).

A certain and easy method of separating ionium from any mixture is to add thorium thereto, if not already present, and then to separate the thorium and subject it to a rigorous purification process by the ordinary methods. This is a general method adopted in radioactivity. In most cases the substance so added is not radioactive and so does not greatly matter, except that it acts as a diluent of the active material. In the present case the activity of the thorium has to be taken into account, but it is usually excessively feeble compared with that of the ionium. It is doubtful if ionium free from thorium has been prepared. Cerium, carefully freed from thorium, has been employed for the separation of ionium, for example, by precipitation with hydrofluoric acid in strongly acid solution. The cerium may then be separated from ionium by the methods used to separate it from thorium. However, there is probably some thorium in all uranium minerals. In Joachimsthal pitchblende the quantity is extremely small, but, in the working up of many tons of mineral, Auer von Welsbach [20] accumulated a few grams of pure thorium containing the ionium of the mineral. The nitric acid solution of pitchblende, after removal of lead, &c., by sulphuric acid, may be precipitated directly with hydrofluoric acid. The fluorides are dissolved by boiling with sulphuric acid and fractionally precipitated by zinc hydroxide paste. The first precipitate contains all the thorium and ionium present.[33]

In the working up of thorium minerals, the whole of the ionium associated with the uranium present remains with the thorium. As practically all thorium minerals contain some uranium, all sources of thorium contain some ionium, the preparations obtained from monazite containing less than those from thorianite. This fact is often overlooked.

The most characteristic feature of the radioactivity of ionium is its power of producing radium steadily with the lapse of time. The radium may be easily separated from the ionium, for example, by precipitating barium sulphate in the solution, but in course of time it continuously reforms. It is usually detected by means of the characteristic radium emanation, the solutions being kept in sealed flasks and the gases accumulating boiled out from time to time and introduced into an air-tight electroscope. Owing to the long period of ionium the amounts of radium so formed are, of course, extremely minute. As a consequence of what has already been said, it follows that all thorium preparations generate radium, to a greater or less extent, according to the amount of uranium in the minerals from which they were derived. Otherwise, ionium, in the radioactive sense, is very similar to polonium, its radioactivity consisting of low-range α-rays only. It possesses the great advantage over polonium for many purposes, in that its radioactivity is constant and does not sensibly decay. It will no doubt find large application where a steady, powerful source of α-radiation is required, as in compensation methods of radioactive measurement and for α-ray standards more active than those of uranium. The proportion changing into radium, even in a lifetime, is so small that it may in most cases be neglected. It is only by refined emanation tests that it can be detected.

The period of ionium remains uncertain. It is certainly very long, at least fifteen times as long as that of radium, if no other long-lived intermediate substances exist in the series. This follows from the failure of the experiments to detect a production of radium from carefully purified uranium preparations. On the other hand, from the intense α-activity of Welsbach's ionium preparation separated from pitchblende, it follows that the period cannot be greater than a million years.[38] From the range of its α-particles, an indirect estimate has given the period as nearly a million years.[14]

RADIUM (RA)

Atomic weight—226.5.
Period of average life—2500 years.
Radiation—α-rays and (β)-rays.
Range of α-rays—3.13 cms. of air.
Penetration power of (β)-*rays*—Half absorbed by 0.022 mm. of aluminium.
Nearest chemical analogue—Barium.
Parent—Ionium.
Disintegration product—Radium emanation.

Occurrence.—All ores containing uranium contain radium. In the great majority of cases the ratios of the quantity of radium to that of uranium is a constant, viz. 3.2×10^{-7}, which is known as the "equilibrium ratio." That is to say, 320 milligrams of radium (element) are present per ton of uranium (element). Certain minerals are exceptional in that the constant of radium is less than this. The most important of these are the autunites (Autun, France, and Guarda, Portugal), a hydrated calcium uranium phosphate $CaO, 2UO_3, P_2O_5, 8H_2O$, for which the ratio varies from 20% to 80% of the equilibrium. Conversely, the only certain case of a mineral containing radium and no uranium is a certain deposit of pyromorphite (Issy l'Evêque, France) found in the neighbourhood of autunite deposits, which are superficially coated with a radiferous layer, doubtless derived from the autunite by the action of percolating water. According to recent researches, it seems that geologically old formations contain a somewhat greater radium ratio than recent ones.[39]

All common rocks and minerals of the earth's crust contain minute amounts of radium, of the order of a few million millionths of the total mass. It is found in natural springs, both hot and cold, many of the spas most celebrated medicinally containing the greatest quantities either of radium or of its emanation. There are few materials in which radium cannot be detected by the emanation test, if sufficiently large quantities are examined, but whether this has any special significance or is merely due to the almost incredible delicacy of the emanation test is not easy to decide. There is no reason to doubt that in all these common materials the three-millionfold greater equilibrium quantity of uranium is also present, although this cannot be detected chemically

Properties.—Radium is the most important of the new radioactive elements. It was discovered by M. and Mme. Curie in the years immediately following Becquerel's first observation of the property of radioactivity of uranium compounds in 1896. Mme. Curie formulated the fundamental principle that radioactivity is an atomic property, and this led her to the examination by chemical analysis of the uranium minerals, the radioactivity of which is several times greater than can be accounted for by the radioactivity of the uranium present. She found the radioactivity was concentrated in the second group with bismuth (polonium) and in the alkaline earth group with barium (radium). Radium possesses all the usual analytical reactions of barium, the sulphate being even less soluble in water than barium sulphate. It is therefore the least soluble sulphate known. Radium is distinguished from barium by the lesser solubility of its chloride and bromide in water or in hydrochloric and hydrobromic acid, and this is the principle of the method adopted in its separation from barium. The physical and chemical properties (heat of solution and of formation, solubility, facility of forming peroxides, stability of the carbonates, &c.) of the various compounds of the alkali and alkaline-earth metals usually show a well-marked graduation in the order Ca, Li, Sr, Ba, Na, K, Rb, Cs, and in this scheme radium comes between barium and sodium.[40] This may prove a valuable analogy in indicating the probable properties of the various radium compounds, which are still for the most part unelucidated.

Spectrum.—Radium compounds colour the bunsen flame a fine carmine tint. The flame and spark spectra exhibit the closest analogy to those of the other alkaline-earths. The strongest line in the photographic region of the spark spectra is in the ultra violet ($\lambda = 381.47\mu\mu$). This line is visible in the spectrum of a barium salt containing only 0.01 per cent. of radium. The relative intensity of the two neighbouring violet lines, of radium 455.33, and of the strongest barium line 455.42, furnishes a good test of the completeness of separation of the elements. Even in the purest fractions of Mme. Curie the barium line was still just visible, while when

0.6 % of barium was added the intensity of the two lines became similar.

Separation and Fractionation of Radium.—The most important operations in the working up of radium-containing materials are the solution of the material, consisting usually of insoluble sulphates, and the separation of the halogen salts of the alkaline-earth group in a pure state, followed by their fractional crystallisation. The first operation is usually effected by vigorous boiling with sodium carbonate solution, filtering and washing free from sulphate. This is the well-known reaction studied dynamically by Guldberg and Waage, whereby an equilibrium is attained between the two pairs of soluble and insoluble sulphates and carbonates. Naturally the greater the excess of sodium carbonate the larger the proportion of insoluble sulphate converted into insoluble carbonate. In this operation it is advisable not to wash at once with water but with sodium carbonate solution until most of the sulphates are removed, as thereby the reconversion of the carbonates back into insoluble sulphates is largely prevented. In dealing with crude materials, for example, the radium-containing residues from pitchblende, it is often advantageous to precede this operation by a similar one, using a sodium hydrate solution containing a little carbonate which dissolves part of the lead and silica present. The carbonates, washed free from sulphates, are treated with pure hydrochloric acid which dissolves the alkaline-earths including radium. From the solution the latter may be precipitated as sulphates by sulphuric acid and reconverted back into carbonates as before, or sometimes more conveniently they may be precipitated directly as chlorides by saturating the solution with hydrogen chloride. This is a very elegant method of great utility in the laboratory, for the most probable impurities, chlorides of lead, iron, calcium, &c., remain in solution and only the barium and radium chloride are precipitated, practically in the pure state, ready for fractionation.

All these tedious and lengthy wet reactions could with advantage be replaced in many cases by simple reduction of the insoluble sulphates to sulphides in a current of coal gas, water gas, or other reducing atmosphere, followed by solution of the product in acids.

The fractionation offers no difficulty, at least in the initial stages when the quantities of materials are large. The chloride is dissolved in boiling water in one vessel, A, to a just saturated solution and left to cool, the mother liquor is drained off into another vessel, B, again concentrated to saturation, while the crystals are dissolved to saturation in fresh boiling water. On cooling, the mother liquor of B is drained into a new vessel, C, while that of A is drained into B, and used to dissolve the crystals. The operation is repeated till perhaps five vessels are obtained. After this the number of fractions is kept constant while, alternately, crystals are withdrawn from the process at the less soluble end and the mother liquor from the other end. The crystals contain practically all the radium and the mother liquor only a small quantity. When the crystals become very small in quantity, through repeated fractionation, gradually increasing quantities of hydrochloric acid are added to the water used to dissolve them, so increasing the volume of the solutions handled and enabling the crystallisation of only a few milligrams of the salt to be conveniently carried out. The fractionation of the bromides is described as more rapid but less regular. The chloride is a much more stable compound than the bromide, and parts with its halogen much less readily on keeping.

Estimation—(1) *Chemical Methods.*—The estimation of the radium in a preparation is accomplished by a variety of methods according to the circumstances. For mixtures of pure radium and barium, salts containing an important proportion of radium in the absence of other elements, complementary to the spectroscopic test described above, is the determination of the chemical equivalent. Mme. Curie precipitated the solution of the chloride with silver nitrate, weighing the silver chloride and recovering the radium from the solution in the soluble form. The chloride of radium must be dried at 150° and weighed quickly, as it is hygroscopic. Sir William Ramsay converts the chloride into bromide by ignition in bromine vapour. The only objection to estimating the radium as sulphate, as in the case of barium, is the difficulty of recovering the radium in soluble form.

The solution of radium sulphate may conveniently be

effected by gentle ignition in a current of illuminating gas, followed by solution of the sulphide produced in hydrochloric acid ; or by boiling with sodium carbonate solution as already described. If x is the equivalent of the mixture of barium and radium found by any of these methods, and A and (100 − A) are the percentages of radium and barium respectively, obviously

$$x = \frac{A}{100} \cdot \frac{226.5}{2} + \frac{100 - A}{100} \cdot \frac{137.4}{2}$$

so that from the value of x, A can readily be deduced.

(2) *γ-Ray Method.*—The simplest and most direct way of estimating the quantity of radium in any preparation is by comparison of the γ-rays.

Provided that a standard preparation of known radium content is available, the percentage of radium in any preparation may, in the absence of mesothorium, readily be estimated by a simple comparison of its γ-radiation with that of the standard preparation. The γ-radiation is so penetrating that the absorption of the rays even in a centimetre thickness of the material is not serious, and hence this method enables substances of very different weights and densities to be compared together, although it is advisable to use as a standard a substance as much as possible like the material being tested in these respects. As the γ-rays, and the same is equally true of the β-rays, do not come directly from radium, but from its products, escape of emanation must be avoided and the preparation must have been prepared at least a month. The only perfectly rigorous method is to seal up the preparations in glass tubes a month prior to testing. But for the rapid estimation of the radium in minerals this precaution can usually be omitted. This method lends itself especially to this kind of test, and is the one invariably employed by the author. A large lead * electroscope, say 20 cm. high by 20 cm. diameter, of wall-thickness 0.4 cm., the windows and exposed parts screened by lead to prevent the entrance of secondary radiation, is employed. The leaf system is the simple single

* Old sheet-lead from the roof of an ancient building is best, as the radio-lead, always present in freshly prepared lead, will have decayed, and the natural leak of the instrument will therefore be less.

leaf system insulated by a sulphur bead. In comparing pitchblendes and other minerals a lump of from 20 to 100 grams according to the richness is simply laid on the top, and the increase of leak compared with that produced by a similar lump of a standard pitchblende, the uranium content of which is known. Thus as an example:

	% U.	Weight.	Leak.
Standard . . .	60	25 g.	15 div. per min.
Specimen tested . .	x	40 g.	10.4 div. per min.

$$x=\frac{25}{40}\times\frac{10.4}{15}\times 60=26\% \text{ U},$$

and the amount of radium is $0.26\times 0.32=0.083$ gram per ton of mineral.

For materials containing only of the order of one per cent. of uranium, larger quantities are necessary, and it is convenient to place them in curved shallow boxes of tinplate around the sides of the electroscope. For the standard of comparison the boxes may be refilled with a mixture of the material being tested with a known small proportion of powdered standard pitchblende. This avoids all uncertainty due to the absorption in the material itself.

The comparison of concentrated radium preparations in sealed tubes can be done readily by the same instruments, the distance of the preparation from the instrument being varied to suit the strength of the preparation. Provided a proper lead electroscope is used no difficulty is to be anticipated from the strong secondary radiation generated by the γ-rays in all surrounding objects, but otherwise it is important that nothing should be moved on the bench during the measurements except the active preparations themselves, and that the observer should sit motionless in the same orientation throughout the actual observations.

In the presence of thorium, the test gives still a reliable measure of the technical value of the mineral. The separable mesothorium contributes per unit about one-fifth to one-sixth of the γ-ray effect contributed per unit by the uranium, whilst the non-separable radio-thorium probably contributes a like amount. The radium separated from thorium-containing uranium minerals contains also the whole of the mesothorium present. The possibility also of the wilful adulteration of radium preparations by means of the less permanent mesothorium must not be forgotten. The γ-ray test gives really in such cases the measure of the penetrating radioactivity separable as radium rather than a measure of the true radium. Thus if a mineral, containing equal percentages of uranium and thorium, appeared by this test to contain 30 milligrams of radium to the ton, the real radium content would be between 25 and 20 milligrams. After the lapse of many years, when the decay of the mesothorium has occurred, the activity would of course sink to the true radium value. In such cases the method of procedure is to take the γ-activity three hours after the preparation has been dissolved in water and evaporated to dryness again. This allows the radium emanation to escape, and after three hours the γ-activity is that due to the mesothorium alone. (See also *Mesothorium*.)

(3) *The Emanation Method.*—A method applicable to solutions containing from 10^{-8} to 10^{-10} gram of radium is to seal up the solution, which must be limpid, and acid, to prevent the precipitation of the radium, in a small distilling flask, and to allow the emanation to accumulate. After one month the equilibrium amount is present, and the quantity then does not alter however long the flask is kept sealed. The emanation is boiled out of the solution in a slow current of air into a gasholder, and the contents of the gasholder are admitted through a calcium chloride tube into an exhaust air-tight electroscope. The leak after three hours from admission is taken, the gold leaf being positively charged. As standards may be employed similar flasks containing small known weights of pitchblende of known uranium content dissolved in nitric acid. If for any reason it is inconvenient to wait a month after sealing before testing, the proportion of the equilibrium amount of emanation accumulating may be taken as $1-e^{-\lambda t}$, where λ has the value 0.0075 $(\text{hour})^{-1}$, and t is the time in hours.[41] (Tables of this function are to be found in *Le Radium*, 1909, 195.)

Radioactivity.—The radioactivity of radium itself, free from the products of its own disintegration, consists of α-rays. In addition there is a feeble and unimportant (β)-radiation. In solution the emanation, which is the first disintegration product, escapes freely, but in the solid state it is retained to a greater or less extent, depending upon the nature of the compound. For the sulphate and dry chloride the retention is fairly complete, for the bromide as much as one-half of that produced may escape, whilst from hydroxides, carbonates, and similar compounds it appears to escape most freely. In a sealed tube of glass, however thin, retention is complete. The penetrating rays come entirely from one of the later products of the emanation (RaC), and therefore the penetrating radiation of a radium preparation varies according to the proportion of the full equilibrium amount of emanation present. A freshly prepared sealed radium preparation attains of this equilibrium amount one-half after four days, three-quarters after eight days, and so on, the equilibrium being practically reached in from three weeks to a month. The sudden removal of the emanation by solution or heating causes no immediate alteration of the penetrating rays, as the emanation itself gives only α-rays. But a gradual decay commences, which is slow for the first half-hour (owing to the existence of the intermediate product RaB), and then is more or less regular ; the activity falling continuously to about half-value every half-hour, and the decay being practically complete in 3 to 4 hours.

As regards the α-radiation, that due to radium itself is about 25 % of the full equilibrium α-radiation of radium and all the products up to radium C. After removal of the emanation there is a rapid fall in the α-radiation, due to the decay of the α-rays of RaA, which is nearly complete in ten minutes, then a pause for half-an-hour, and then a nearly exponential decay to a 25 % minimum much at the same rate as for the penetrating rays. When the radium emanation is introduced into an electroscope the converse of these changes occurs, namely, a rapid rise of activity for ten minutes and then a more gradual rise, lasting about three hours, to a maximum, followed by the slow decay to half-value every four days characteristic of the emanation itself. If the latter is at any time blown out and removed, there is again a rapid decay of activity during the first ten minutes, followed by a more gradual one almost to zero in about three hours. The existence of this most characteristic set of changes of radioactivity affords an infallible qualitative test for the presence of radium by means of its emanation in any quantity, however minute.

The measurements of the α-rays afford the only way of following the progress of concentration during the fractionation of a radium-barium salt, for when freshly prepared from its solution the crystals possess about 25 % of the equilibrium α-activity, but no β- or γ-activity. Mesothorium does not give α-rays, and therefore its presence does not interfere in this case. The difficulties about this kind of measurement centre in the intense ionisation produced by the α-rays from even minute amounts of radium, and the impossibility of cutting it down satisfactorily by screens or by removing the preparation to a distance from the electroscope. It can be overcome by increasing the capacity of the electroscope by means of a condenser, but then the danger is that the potential gradient, due to the charge on the leaf, is insufficient to saturate the gas. With a sufficiently high source of potential, a separate ionisation chamber, and a quadrant electrometer many accurate methods can be devised, but these requisites, especially the first, are not always to be found in a chemical laboratory. The simplest mode of procedure in this case would be to introduce evaporated films of the salt of definite area into an air-tight electroscope, and to exhaust the latter to a definite low pressure, measured by a barometer. At low pressures the ionisation is reduced proportionally to the pressure, and the "saturation" difficulty is avoided.

Naturally the greatest precautions have to be taken in working with bare radium preparations to avoid contamination of the measuring instruments. Their introduction into the same room as the instruments is apt to spoil the latter permanently. Not only is there the ordinary contamination due to the escape of emanation and the coating of all the walls and objects of the room with films of radio-lead and polonium, which is a well-recognised danger, and which must be guarded against by carrying on chemical operations with radium as remote as possible from the instrument room, into which only sealed tubes of radium may be taken. But there is also an actual species of volatilisation from bare films of radium exposed in the open air, which possibly is a kind of recoil effect, and which results in the coating of all the objects of the room with particles of radium itself.[42] Once this has occurred, delicate instruments such as electroscopes are useless, and must be replaced by less sensitive ones, which, as already explained, are less simple, as they require the use of high potentials, with separate ionisation chambers and measuring instruments.

RADIUM EMANATION

Atomic weight—222.5
Density—111.25 ($H=1$).
Period of average life—5.57 days.
Period of half-change—3.86 days.
Radioactive constant—0.00751 $(\text{hour})^{-1}$.
Radiation—α-rays.
Range of α-rays—3.94 cm. of air.
Nearest chemical analogue—Xenon.
Parent—Radium.
Disintegration product—Radium A.

The radium emanation, on account of its comparatively long period and the ease with which it may be separated from any material and from its parent and products, is by far the most completely investigated of the shorter-lived radio-elements. Chemically all three emanations are distinguished by that complete absence of combining power characteristic of the gases helium, neon, argon, krypton, and xenon, so that they are not chemically absorbed by any reagent. On the other hand, they are all condensed at liquid-air temperature. In the case of the radium emanation, the condensation and volatilisation occur with remarkable sharpness usually between the temperature of $-152°$ and $-154°$ C., but the phenomenon depends somewhat on the conditions and the quantity of emanation.

If a radium solution is kept for some time in a closed flask the emanation steadily accumulates with time, half of the equilibrium amount forming in four days, three-quarters in eight days, and so on, equilibrium being practically reached in three or four weeks. In addition, hydrogen and oxygen are formed from the solution by the radioactive decomposition of the water. On exploding the gases, a slight excess of hydrogen remains with the emanation. This may be admitted to a vacuous bulb immersed in liquid air. The emanation is condensed, and the hydrogen, together with any helium produced from the emanation, can be pumped away. As the emanation in presence of oxygen compounds exerts a powerful oxidising action on organic compounds, for example, tap grease, the emanation so obtained is usually contaminated with oxides of carbon, and may be purified by long contact with freshly ignited lime or baryta. The volume of emanation, at N.T.P., in equilibrium with 1 gram of radium, is almost exactly 0.585 cub. mm.[43] This latter value is the theoretical volume on the assumption that the emanation molecule consists of single atoms, so that the agreement between the theory and experiment directly proves the monatomic character of the molecule. The spectrum consists of a large number of bright lines, and is similar to that of xenon, its nearest analogue.[44] On account of its relationship with the argon group the name "niton" has been suggested, but has not been generally adopted on account of the obvious advantages of its original name in expressing its radioactive relationships and of the disadvantage in proposing a new name for one only of the three emanations known. Its atomic weight has been recently determined from its gas-density, both by Bunsen's method, using its rate of effusion through a perforation in a plate,[45] and directly by weighing on a micro-balance of quartz sensitive to a millionth of a milligram.[46] Although in small quantities its condensation and volatilisation is remarkably sharp, in larger quantities it possesses an appreciable vapour pressure, the curve of which has been determined, from the critical temperature 104.5° and pressure 47,450 mm. to its freezing-point, −71°, at which its vapour pressure is 500 mm. The liquid emanation under the microscope is described as colourless and transparent, whereas the solid is opaque, and glows with great brilliancy like a tiny arc-lamp with steel-blue colour, which changes to brilliant orange-red at a lower temperature.[47]

The production of helium in spectroscopic quantity was first observed for the radium emanation on about 30 mgm. of radium bromide.[48] In later experiments the emanation has been enclosed in a very thin-walled capillary tube, thin enough to allow the α-rays to pass through, but perfectly impervious to ordinary helium. The production of helium was detected outside the tube, proving directly the identity of the α-particles and helium. The α-particles have sufficient velocity to penetrate into ordinary materials to a depth of a few hundredths of a millimetre, and the helium produced from the emanation is largely driven into the walls of the containing vessel, and slowly diffuses out again in the course of time.[11]

The peculiarities of the radioactivity of the emanation consequent upon the production of the non-volatile "active deposit" have already been fully discussed under the emanation method of estimating radium (p. 50), where the differences in the ease with which the emanation escapes from various radium compounds have also been dealt with. It may be added that wherever possible it is best to obtain the emanation from a radium preparation by dissolving it, but if this is not practicable the great part may usually be driven out by strongly heating.

Cocoanut charcoal has the power of absorbing the emanation at the ordinary temperature, and may be used for this purpose when liquid air is not available. The emanation is again given out completely when the charcoal is strongly heated, but it is also evolved quite rapidly at ordinary temperature. If the charcoal that has absorbed emanation is exposed to the open air for a few hours, by far the greater portion of the emanation escapes. Other solid substances, as meerschaum, finely divided metals, and even ordinary metal surfaces have the power of absorbing the emanation to some extent. This is a possible source of error in certain experiments, for example, when the subsequent activity of metal plates which have been exposed to the emanation is being studied.

The radium emanation is distinctly soluble in water, the coefficient of solubility being about 0.3 at ordinary temperature, and 0.12 at 80°. It is much more soluble than this in all organic liquids, except glycerine. In petroleum and toluene, the solubility coefficients are 9.5 and 11.7 at the ordinary temperature. It is less soluble in salt solutions than in pure water.[49]

RADIUM "ACTIVE DEPOSIT"

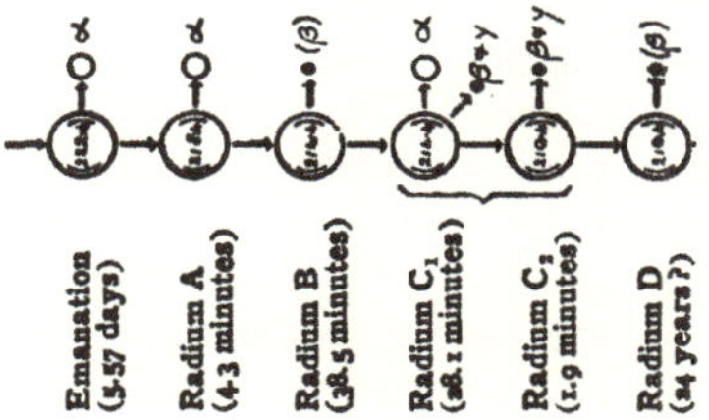

	RaA	RaB	RaC_1	RaC_2
Estimated atomic weight . . .	218.5	214.5	214.5	?
Period of average life (minutes) .	4.3	38.5	28.1	1.90
Period of half-change (minutes) .	3.0	26.7	19.5	1.38
Radiation	α-	(β)-	α-, β- & γ-	β- & γ-
Range of α-rays (cm. of air) . .	4.50	...	6.57	...
Penetration power of β-rays (mm. of aluminium for half-absorption) .	...	0.09	About 0.5	?
Penetration power of γ-rays (cm. of lead for half-absorption) . .	...	...	1.38	

The radium "active deposit of rapid change," so called to distinguish it from its products, radium D to radium F, which are more permanent, and are sometimes therefore spoken of as the radium "active deposit of slow change," comprises a group of four successive products. The first member, radium A, is much shorter-lived than the two succeeding members, and gives α-rays on disintegration. In consequence all experiments on radium A must be carried out within a very short time of its production from the emanation. If a negatively charged wire is exposed for a few seconds only in a relatively large quantity of radium emanation, the activity of the wire at first is almost entirely due to the α-rays of radium A, and these decay to a very small value in a few minutes after withdrawal of the wire from the emanation, the activity of the subsequent products formed being, for such short exposures, usually negligible. After twenty to thirty minutes' exposure to the emanation, the quantity of radium A attains equilibrium, and does not further increase. But the amounts of radium B and C on the wire go on increasing with the time of exposure up to three or four hours. Obviously the shorter the time of exposure the purer is the radium A deposited. Conversely with a wire which has been exposed any length of time to the emanation, no radium A remains after twenty to thirty minutes from withdrawal, the activity at this stage being due to radium B and C. Hence, without chemical treatment, radium A alone or radium B and C together may be obtained at will by suitable choice of the time of exposure to the emanation and of the time elapsing after withdrawal.

The decay of the active deposit after withdrawal from the emanation naturally varies very much according to the time of exposure. There is always a rapid initial decay of the α-rays due to the change of radium A, which is the more marked the shorter the time of exposure to the emanation. This is followed by a period of very little change for about half-an-hour, as the decay of radium C is more or less balanced by its production from the practically rayless radium B. Then follows a continuous decay to nearly zero in about three or four hours, with a half-period intermediate between that of radium B and radium C. The penetrating rays are a measure of the quantity of radium C alone present, and therefore do not show any initial decay due to radium A, but rather an initial increase, or lag in the rate of decay, according to the time of exposure, as radium C is being produced by the changes of radium A and B.

On heating the wire carrying the active deposit, the volatility of the products is very different.[50] Radium B is strongly volatilised at about 600°, whilst radium C requires a much higher temperature, and does not commence to volatilise till about 1100°. As regards radium A, it often appears volatile under these conditions, but this is probably a complex "recoil" phenomena. Its true volatilisation temperature is probably between 800° and 900°, for if a wire, maintained at the latter temperature, is exposed to the emanation it does not become active, however charged.[51] The following volatilisation points are given: radium A, 900°; radium B, 600°; radium C, 1100°. A practical method for obtaining radium C by itself is to allow the radium A on the wire to decay, and then to expose the wire for some minutes to a temperature somewhat above 630°. Radium B distils away, leaving radium C on the wire, the total activity of which then decays exponentially with the half-period of 19.5 minutes. The other method mostly employed for this purpose is to dissolve the active deposit off the wire in boiling dilute hydrochloric acid, and to shake up the solution with finely divided metallic nickel, or to plunge into it a copper or nickel plate. Radium C alone deposits on the metal, leaving radium B in solution.[52] The radium B, so left, of course immediately starts to produce a fresh amount of radium C by its own change. Itself it produces only a feeble (β)-radiation, but it is characterised by its power of generating with the lapse of time the powerful α-, β-, and γ-radiation of radium C. Similarly, by electrolysis of the solution of the active deposit with a feeble current and large electrodes, radium C alone is deposited on the cathode.

Radium B may readily be obtained from radium A by recoil.[53] At one time it was thought that radium B was volatile at the ordinary temperature, but its "volatility" is confined to the nascent state, that is, to the radium B actually being produced from radium A. After all the latter has disappeared, the radium B does not volatilise below a red-heat. At ordinary temperatures with radium A present, the radium B at the moment of its formation is largely projected from the plate or wire coated with the active deposit, by recoil of the radium B molecule when the radium A α-particle is expelled. The collection of such recoiled products on surfaces is most favourable in a vacuum, with the receiving surface kept negatively charged.

Radium C has recently been shown to be complex, radium C_1, the member with the half-period of 19.5 minutes, giving a product radium C_2 with half-period of 1.38 minutes only.[54] This change has many features of interest, and is not yet completely understood. So far only very small quantities of the short-lived body have been separated. Radium C_1 gives α-rays, whilst radium C_2 gives β- and γ-rays, but not α-rays. It is probable, however, that radium C_1 gives also β- and γ-rays, and that only a small part of the total penetrating radiation is due to radium C_2. Radium C_2 is obtained from radium C_1 by recoil, but in such small amount that it is probable that the change of radium C_1 giving rise to radium C_2 is not an α-ray change, but a β-ray change, the recoil from which is very much more feeble. But radium D has been proved to recoil from radium C_1 in such relatively large amount that there can be no question that radium D is the direct product of the α-ray change of radium C_1. Hence radium C_1 appears to suffer a multiple disintegration, some of its atoms giving β-rays and radium C_2, others, and by far the larger proportion, giving α-rays and radium D. The interest attaching to these changes is that possibly they may clear up the mystery surrounding the parentage of actinium, to account for which, as already explained, a multiple disintegration has been suggested. Many other phenomena suggested the complexity of radium C which has so recently been established, but little can yet be said of radium C_2 beyond the fact of its existence and its period of life. Its β- and γ-rays seem identical with those given by the active deposit before its separation.

RADIO-LEAD OR RADIUM D

Estimated atomic weight—210.5.
Period of average life—24 years (?).
Radiation—(β)-rays.
Nearest chemical analogue—Lead.
Parent—Radium C_1.
Disintegration product—Radium E.

This element has not yet been separated or concentrated from lead, which, so far as is known, it resembles perfectly in all its chemical reactions. As lead is an invariable constituent of all uranium minerals (except autunite), and in most is present in notable quantity, the lead separated from such minerals always contains the radio-lead associated with the uranium. Conversely, common commercial lead contains traces of the element which causes it to be distinctly more radioactive than most other metals. For making instruments, a very old lead (such as from an old roof) should if possible be employed, as in this the radioactive constituent will largely have decayed. Although itself rayless, except for an excessively feeble and unimportant (β)-radiation, its product, radium E, gives β-rays, and as the latter is of short period, the β-radiation is rapidly produced. More slowly an α-radiation makes its appearance, due to the production of radium F (polonium). In consequence the lead separated from uranium minerals shows, a few days after separation, a marked β-activity, and in the course of a few months an α-activity also. By purification the radio-lead is very easily obtained again inactive, but only to regenerate its activity in course of time. Owing to the large amount of lead in uranium minerals, radium D as yet is technically of little value, for no means of concentrating the activity exists. If it could be concentrated it would for many purposes be as valuable as radium itself.

Another source of radio-lead, however, is old radium. The period of radio-lead, although long, is still short enough to cause its production from radium to be easily detectable. After the decay of the active deposit of radium is complete, an extremely feeble activity is left, which then increases, the β-radiation for a few weeks and the α-radiation for a couple of years. This is due to radium D, which, after formation, produces first radium E and then radium F. In old radium preparations which have been kept in the solid state, or under conditions where the emanation is retained, the radio-lead steadily accumulates, and may be separated with the radium, from the polonium and radium E, by precipitation with sulphuric acid. From the radium it may be separated by adding a little lead and then separating it by any convenient method.

However, if a preparation of radio-lead is required as concentrated and pure as possible, the simplest method of, procedure is to remove the emanation periodically from a radium solution kept in an air-tight vessel, and to introduce it into a closed glass flask. As the emanation decays, radio-lead is formed and accumulates as an invisible film on the walls of the flask. Tubes that have contained extremely concentrated emanation from large quantities of radium show a sub-metallic deposit on the glass, which perhaps may be due to visible amounts of pure radio-lead and its products.

RADIUM E OR RADIUM E_1 AND E_2

Estimated atomic weight—210.5.
Period of average life—7.25 days.
Period of half-change—5.10 days.
Radiation—β- and γ-rays.
Penetrating power—β-rays half absorbed by 0.162 mm. of aluminium.
Nearest chemical analogue—Unknown.
Parent—Radium D.
Disintegration product—Polonium (Radium F).

This short-lived member gives out β-rays, and stands between a long-lived rayless element and a less long-lived α-ray producing element. It is therefore completely analogous to mesothorium 2 of the thorium series. It cannot be considered as yet very completely known, for whereas at one time it was regarded as complex, consisting of a rayless element, radium E_1, of half-period 6.5 days, which produced the β-ray giving element, radium E_2, with half-period 4.8 days, later evidence has shown that the radiations come from the direct product of radium D. Thus the radium D deposited from a sufficiently large quantity of radium emanation stored within a sealed glass tube for a few hours, develops the β-radiation of radium E normally and regularly with the period given above, whereas, if a rayless product intervened, the curve would be complex, and the growth of the β-rays would be retarded. Moreover, the β-rays decay regularly with the same period when, from a preparation of radio-lead, radium D is removed by precipitating barium sulphate in the solution.[55] On electrolysing a solution of acetate of lead, containing radio-lead and its products, with gradually increasing current densities, the polonium is first deposited on the cathode (at 4 micro-amperes per $cm.^2$), then polonium and radium E separate (at 10 micro-amperes), while with heavier currents all the products are deposited. This is an example of a generalisation which holds good with the exception of thorium D. The successive products of a disintegration series become electro-chemically more "noble," and are deposited the more easily as we pass through the series. They also, as a general rule, become more volatile and more readily soluble in acids.

As they have been described, radium E_1 is volatile at a red heat, whilst radium E_2 is not volatile. The first substance is said to be soluble in hot acetic acid, while the second is not soluble, when an electrolytically deposited film is examined.[56]

The β-rays of radium E are homogeneous, and are absorbed exponentially by aluminium, the coefficient of absorption being 44. This is about three times greater than that of the β-rays of uranium X, so that the radiation is comparatively feebly penetrating. The γ-rays accompanying the β-rays are extremely feeble, and possess comparatively little penetrating power. They therefore do not interfere at all with γ-ray measurements of the strength of minerals and of radium preparations when the rays are, as is customary, first passed through 1 cm. of lead. As the β-radiation of radium E decays, an α-radiation due to radium F makes its appearance, but as the period of the latter is long compared with that of its parent, the α-radiation so produced is not very pronounced. In turn, of course, it decays completely with the period of radium F.

POLONIUM. (Radium F)

Estimated atomic weight—210.5.
Period of average life—202 days.
Period of half-change—140 days.
Radiation—α-rays.
Range of α-rays—3.58 cm. of air.
Parent—Radium E.
Disintegration product—Unknown, probably lead.
Nearest chemical analogues—Bismuth, tellurium.

Owing to its position, being intermediate between the long-lived and short-lived radio-elements, two possible sources of polonium are available. It may be obtained from uranium minerals, being separated analytically with sulphuretted hydrogen in group II. and resembling bismuth most closely of the members of this group. Or, it may be obtained from old preparations of radio-lead or radium. The quantity of polonium in radio-lead increases to a maximum, which is reached in about two years, and then steadily diminishes as the parent decays. In old radium, the emanation of which is not allowed to escape, the quantity of polonium must steadily increase for thirty years or longer, and owing to the intermediate existence of radio-lead, a comparatively long-lived member, the production for the first ten years or more is small. Since radio-lead is a by-product of no direct value, it should furnish the cheapest and most convenient source of polonium. On precipitation with sulphuric acid the radium E and radium F remain in the solution, and after a few weeks the former has completely disintegrated, leaving the latter by itself. On electrolysing a solution of lead acetate containing radio-lead and its products, with a current of four micro-amperes per sq. cm. of cathode surface, radium F alone is deposited. With 10 micro-amperes per sq. cm. both radium F and radium E are deposited, and with 100 micro-amperes radium D and lead also are separated.

If the ordinary radium-containing residues of pitch-blende are digested with hydrochloric acid, part of the contained polonium is dissolved, and may be precipitated with sulphuretted hydrogen. From the acid solutions obtained in the working of the residues for radium, polonium may also be similarly obtained. About 3 kilograms of bismuth oxychloride are obtained from 1 ton of Joachimsthal pitchblende. The polonium may be separated (1) by fractional precipitation from solutions made very acid with hydrochloric acid, the polonium being enriched in the precipitate; (2) by sublimation in vacuo, the polonium being the more volatile; (3) by fractional precipitation of the basic nitrate with water, the precipitate being enriched (Mme. Curie). Very elegant methods have been devised by Marckwald, who proposed the name *Radio-Tellurium* for the substance before its identity with polonium was established by observations on its period. By immersing a plate of bismuth, silver, copper, &c., in the hydrochloric acid solution, the polonium is practically completely precipitated. In the same solution stannous chloride produces a small black precipitate consisting mainly of tellurium and containing nearly all the polonium. If this is dissolved in not too acid solution and hydrazin hydrate added, the tellurium is precipitated and the polonium remains dissolved, and may be precipitated by stannous chloride. In this way 4 milligrams was obtained from 2 tons of pitchblende. Naturally it was intensely radioactive. A hundredth of a milligram sufficed to render a zinc sulphide screen plainly visible in the dark to a large audience.[57] The theoretical quantity of polonium in minerals is 1 milligram for every 14 tons of uranium (element), assuming that no radium emanation escapes from the mineral. Latterly Mme. Curie, separating the polonium from several tons of pitchblende, obtained as final product 2 milligrams, which was estimated to consist of polonium to the extent of 5 %. It showed several new lines in its spectrum, which are being periodically examined as the polonium disintegrates, to see if they disappear and give place to the lines of lead.[58]

The exponential decay of the activity of polonium is perfectly regular and complete, about one-half per cent. of the activity decaying daily, the half-value being reached in twenty weeks. Like ionium, it gives only α-rays of comparatively low range, no effect being produced through a single thin sheet of paper, even with the most intense preparations. It has proved an extremely useful radioactive substance, but for most purposes ionium would now be preferred on account of its greater permanence.

THORIUM (Th)

Atomic weight—232.5.
Period of average life—Unknown. Probably about five times greater than that of uranium.
Radiation—α-rays (? may be rayless).
Range of α-rays—3.5 cm. of air.
Disintegration product—Mesothorium 1.

The chief sources of thorium at the present time are (1) monazite sand from Brazil and N. and S. Carolina, U.S.A., from which practically all the thorium used technically is obtained ; it consists of 60 or 70 per cent. of monazite in fine grains with other sands, composed of granite, rutile, zircon, magnetic and titaniferous iron stone, not attacked by acids ; (2) thorianite, a rare mineral from Ceylon, which contains as much as 60 to 70 per cent. of ThO_2 in a form soluble in moderately strong nitric acid and from 10 to 20 per cent. of U_3O_8. Formerly thorium was obtained principally from the silicates, thorite and orangeite, found in Norway, but these sources are nearly exhausted. Monazite is mainly a phosphate of the rare-earths, cerium, lanthanum, præsodymium, and neodymium, with variable amounts of thorium. Usually the commercial sand contains 4 to 5 per cent. of ThO_2, having been concentrated by washing to this percentage. In addition small quantities of iron oxide, lime, alumina, and silica are present.

In the first stage of the technical treatment of this sand, it is heated with twice its weight of sulphuric acid. The cold mass is dissolved in water, and left to settle. The solution is then fractionally precipitated with magnesia, the thorium being concentrated mainly in the first fractions precipitated. The commonest and most useful reagent for precipitating the rare-earths from a solution containing common earths such as alumina, iron, &c., is oxalic acid. Now thorium oxalate is of all the rare-earth oxalates the least soluble in acids, so that by working in fairly strong nitric acid solution thorium oxalate may often be precipitated and separated at least partially from the other rare-earths and from calcium. The same is true of the rare-earth phosphates, that of thorium being one of the most insoluble in dilute acids. On the same principle, thorium is often precipitated by weak bases, such as the substituted ammonias, for example, dimethylamine, while zirconium, &c., remain dissolved. The potassium salt of hydrazoic acid, KN_3, precipitates thorium hydroxide only from mixtures of thorium and cerium on boiling. The same separation may be effected by means of sodium thiosulphate on boiling, thorium alone being separated, as hydroxide. This ready hydrolysis of weak thorium salts is characteristic of the element. The oxalates of thorium and zirconium alone of the rare-earths are soluble in ammonium oxalate, and on strongly acidifying the solution the former alone is reprecipitated. The solution of the oxalate of thorium and its conversion into soluble salts may be effected by means of concentrated ammonium or sodium carbonate and precipitation of the concentrated solution as thorium hydroxide with strong ammonium or sodium hydrate. Thorium is distinguished from the yttrium group of the rare-earths by its power of forming a double sulphate with potassium sulphate, insoluble in excess of the latter reagent, and so may be separated from a mixture of the sulphates by saturating the solution with potassium sulphate. Alike in the old, now obsolete, as in the present technical methods of purifying thorium the peculiar solubility relations of thorium sulphate in water have been largely applied. The older method consisted in volatilising the excess of sulphuric acid from the material being treated, and in dissolving the anhydrous sulphates in ice-cold water—a tedious operation—and in heating the solution till the hydrated thorium sulphate was precipitated. The latter was then dehydrated at 300° to 400°, and the process repeated. In present practice the sulphuric acid is always kept in great excess in the initial treatment of the mineral, but the sulphate method may be employed at the final stage of manufacture as follows. The thorium hydroxide is dissolved in hydrochloric acid, so that the solution contains not more than 30 % ThO_2, and sulphuric acid is added to the extent of a half per cent. more than the equivalent quantity, the temperature being kept low, and in any case below 40° as a maximum. Under these conditions the hydrate $Th(SO_4)_2 8H_2O$ is precipitated, departure from the conditions causing the separation of the tetrahydrate, which is in every way less easily manipulated. The precipitated sulphate is reconverted into hydroxide, and the process repeated as often as necessary to remove all impurities.[58A]

Thorium forms a curious compound with acetonyl acetone, $Th(C_6H_7O_2)_4$, which is soluble in chloroform and alcohol, and can be distilled in a vacuum, and so can advantageously be employed for the purification and separation of the element.

It may be mentioned that in the fusion of refractory minerals, for example, with sodium carbonate, the thorium, if present, is converted into the highly insoluble oxide, ThO_2, and its presence is apt to be overlooked.

Thanks largely to the thorium industry, in which a product unusually pure is essential, there exist, therefore, a great variety of exceedingly good and sharp methods for the separation and purification of thorium, and it must be understood that ionium, if present, and radio-thorium always remain unseparated from thorium in these processes as far as they have been examined.

A glance at the disintegration series shows that radio-thorium and its products are responsible for by far the larger part of the radioactivity of the whole series, just as radium is for that of the uranium series. The three bodies preceding radio-thorium in the series yield only a very small proportion, if any, of the α-rays, and about half of the β- and γ-rays. Thorium, free from radio-thorium, has not yet been prepared, and although this is now theoretically possible, the operation would require many years. For this reason, and also on account of the ionium always present from the uranium of the original mineral, it is difficult to decide definitely whether thorium, free from its products, does or does not give α-rays. It is commonly regarded as giving an α-radiation, but this is certainly an unusually small proportion of the total α-radiation in equilibrium, and may not inconceivably have been due to ionium in the preparations studied. However, the most recent work is against this view, as the range of the α-rays, from a thorium preparation in which the radio-thorium has largely decayed, is distinctly greater than that of the α-rays of ionium.[14]

The non-separability of radio-thorium and thorium results in all thorium preparations in the ordinary state yielding α-rays of the same order of intensity as that of uranium preparations, but more penetrating (those of thorium C being the most penetrating known), and β-rays about one-tenth as intense, and less penetrating than those of uranium. The γ-rays, like the β-rays, are relatively feeble, but those of thorium C are the most penetrating known, while those of mesothorium 2 are only slightly less penetrating than the radium C γ-rays. Freshly prepared from minerals, and retaining therefore the full equilibrium amount of radio-thorium, the activity of thorium salts attains a maximum after one month (due to the accumulation of thorium X), and then steadily decays. Thorium X and mesothorium 1 resemble radium chemically, and are always separated completely in the processes of manufacture. The absence of the mesothorium initially causes the α-rays due to radio-thorium and its products to decay to a minimum, which is about one-half of the maximum value, in about four and a half years, when further decay is just balanced by fresh radio-thorium resulting from the now partially regenerated mesothorium. The activity must then increase again for several decades until the maximum value is regained. Thus the radioactivity is a complex function of the age of the thorium preparation, and for this reason alone thorium is entirely unsuited for use in the preparation of radioactive standards. But in another respect the radioactivity of this element is quite different from that of uranium. All thorium preparations give out, according to their nature, more or less of the thorium emanation, the volatile product of thorium X with period of half-change of about one minute. If the electroscope is air-tight, this emanation goes on accumulating for some ten minutes after a thorium preparation is placed inside, so that the activity on this account rises over this period, but may be reduced again to its first value by blowing out the air of the instrument. The proportion due to the emanation is the greater the greater the mass of the preparation, whereas the proportion due to the α-rays of the solid substance is fixed only by its surface. In consequence, the effect of the emanation is more marked with thick than with thin layers. After being exposed to the emanation for some time, the inside walls of the electroscope become radioactive through the "active deposit," which, once formed, takes some days completely to decay. Boiling solutions of thorium probably allow the emanation to escape practically without any decay of its activity. Cold solutions through which the air is bubbled, and most solid preparations of thorium hydroxide and carbonate, have about the same "emanating power," which is about one-half that of the boiling solutions. This means that in these cases, on the average, a molecule of emanation takes about one minute from the time of formation before it enters the gas phase.

Commercial thorium oxide has an emanating power about one-third that of the hydroxide, but by ignition the emanating power is permanently reduced, the more the higher the temperature employed, until it reaches a limit at about one-tenth of that of unignited oxide. Solid thorium nitrate has even less emanating power than this. In these cases the molecules of emanation diffuse out of the solid so slowly that the vast majority disintegrate within the solid itself and never escape. Naturally the radioactivity of the solid non-emanating compounds is higher on this account than that of those from which the emanation escapes easily. On dissolving these preparations the differences in their emanating power disappear.

MESOTHORIUM 1

Period of average life—7.9 years.
Period of half-change—5.5 years.
Radiation—Rayless.
Nearest chemical analogue—Radium.
Parent—Thorium.
Disintegration product—Mesothorium 2.

This important substance was discovered some years ago,[59] but its chemistry and method of preparation were, until recently, technical secrets. Owing to the enormous quantity of thorium required for the gas-mantle industry, in which mesothorium is a by-product, and to the growing scarcity of uranium minerals, it is probable that mesothorium, in spite of its comparatively short period of average life, will be more largely manufactured than radium in the future. Although itself rayless, its first product, which is short-lived and therefore rapidly formed after it has been separated, gives a powerful β- and γ-radiation. Owing to the subsequent production of radio-thorium and its products, mesothorium preparations acquire an α-radiation with time and the β- and γ-radiation is augmented. A maximum is reached in 4.6 years, when the β- and γ-radiation should be about twice that due to mesothorium 2 alone, and the total radioactivity should represent considerably more than one-half of the activity of the thorium in the mineral from which it was derived. After that time the activity, so far as it is due to mesothorium, will decay exponentially to zero, the half-value being reached in 5.5 years, or about 10 years from preparation.

The chemistry of mesothorium 1 can, like that of ionium, radio-thorium, &c., be accurately described in a sentence. It is absolutely identical in chemical nature with radium and cannot be separated therefrom.[60] In consequence, all radium separated from uranium minerals containing thorium, contains also the mesothorium in the mineral, and all preparations of the latter substance contain the radium present in the mineral from which they are derived. For a mineral containing equal proportions of uranium and thorium the proportion of the γ-rays, which are the ones usually used for measurement, contributed by the radium is some five or six times as great as that due to the mesothorium, when measured through 1 cm. of lead. Fractional crystallisation of the chlorides does not effect the slightest change in the ratio of the two active constituents. The methods employed in the extraction of radium from the residues of pitchblende may be suitably adapted to the working up of the residues from monazite sand. A little barium sulphate may be added to the monazite sand during its treatment with sulphuric acid, and the mesothorium is then separated with the insoluble material left after treatment of the product with water.

The period of mesothorium is several hundred times shorter than that of radium, so that in a preparation having its activity equally distributed between the two constituents, the mass of the chemically identical radium must be many hundred times that of the mesothorium. This fact seems to interpose an impenetrable barrier to the further study of the spectrum, atomic weight, &c., of the element. Even when prepared from a mineral like Ceylon thorite, which has only a very small proportion of uranium, the radium must exceed the mesothorium in absolute quantity, though its radioactivity might be quite negligible by comparison. The preparations prepared technically from monazite may be fractionated till free from inactive matter, and then are stated to be four times as active as pure radium compounds, the activity due to the mesothorium being three times as great as that due to the radium. The composition of such preparations is estimated as 1 % mesothorium to 99 % radium.[61]

MESOTHORIUM 2

Period of average life—8.9 hours.
Period of half-change—6.2 hours.
Radiation—β- and γ-rays.
Penetration power of rays—Half-absorbed, β-rays in from 0.34 to 0.18 of aluminium, γ-rays in 1.1 cm. of lead, or 1.85 cm. of copper.
Nearest chemical analogue—Unknown.
Parent—Mesothorium 1.
Disintegration product—Radio-thorium (?).

Mesothorium 1 behaves like an alkaline-earth, whilst mesothorium 2 behaves like an earth or rare-earth. The latter is in consequence readily separated from the former by adding ammonia to the solution. A trace of some other inactive material precipitable by ammonia must be present, naturally, to serve as a nucleus in the filtration of the precipitate, and a small trace of zirconium is usually employed. The zirconium brings down also any radio-thorium present, and also the thorium X.[62] However, by repeating the process at an interval, say, of one or two days, in which time the mesothorium 2 is to a great extent regenerated, the first precipitate alone contains much radio-thorium, the mesothorium 2 regenerated in the interval being under these circumstances the more free from radio-thorium the shorter the interval. But since radio-thorium gives α-rays, whereas the mesothorium 2 gives only β-rays, some α-radiation is always initially shown by the preparations, and perhaps for this reason the growth of radio-thorium from mesothorium 2 has not yet been directly observed. The period of the former being many thousand times greater than that of the latter, the *growth* of α-rays due to radio-thorium must be, in any case, very small. The β-rays of mesothorium 2 decay regularly with the period given, the decay being practically complete in two or three days. The β-rays are of moderate penetrating power, but are apparently heterogeneous, the absorption coefficient for aluminium increasing from 20.2 to 38.5 $[cm.]^{-1}$ as absorption proceeds. The γ-rays are, relatively to the β-rays, about one-half as powerful as those of radium, and are rather less penetrating, especially in the case of lead. For other metals the difference is not so marked. In this the γ-rays resemble rather those of uranium X. A careful comparison of the absorption of the γ-rays through lead with those from pure radium should reveal the adulteration of radium by mesothorium without the necessity of opening the tube containing the preparation.[68]

RADIO-THORIUM

Period of average life—1063 days (2.91 years) (?).
Period of half-change—737 days (2.02 years) (?).
Radiation—α-rays.
Range of α-rays—3.9 cm. of air.
Nearest chemical analogue—Thorium.
Parent—Mesothorium 2.
Disintegration product—Thorium X.

This radio-element, which resembles perfectly its original parent in chemical nature, would have remained unknowable but for the existence of the intermediate substance mesothorium. The latter can easily be separated both from thorium and from radio-thorium. Hence, after separating it from a thorium mineral, mesothorium is left to itself to produce radio-thorium, which is then separated by adding to the solution a trace of thorium, or, if that is objectionable, of zirconium, and precipitating this out with ammonia. The precipitate retains the radio-thorium and the solution the mesothorium. This is the only known way in which radio-thorium can be prepared. The earliest preparations undoubtedly were prepared in this way unwittingly, mesothorium not then having been discovered. The period has been determined by direct measurements of the rate of decay. Doubt, however, has been cast on the accuracy of this determination, and a recent estimate makes the half-period of radio-thorium only a few months.[63A] When prepared in the manner described the radio-thorium at first retains the mesothorium 2, which decays completely in the course of two or three days. In the meantime thorium X is being produced, which causes an increase of the activity over a period of three or four weeks. The emanating power of the preparation grows with the growth of thorium X. After a month the total activity and emanating power decays to zero. When required for experiments on the thorium emanation, for which radio-thorium is pre-eminently suitable, the thorium hydroxide precipitated from the mesothorium solution should be kept in the moist state and not allowed to dry. With such a preparation all the experiments with the emanation referred to under *actinium* may be repeated with most beautiful and striking results. A puff of air sent through the tube containing the preparation on to the coated surface of a glass zinc sulphide screen, or better, between two such screens placed close together with films inward, gives the most beautiful result, and the decay and reproduction of the emanation with lapse of time can thus be demonstrated easily even to a large audience. From being a phenomenon only to be studied by delicate electric methods, the thorium emanation can now, thanks to the existence of radio-thorium, be shown on the same scale as the emanation of radium, and for illustrating the principles of radioactive change is even more suitable on account of its shorter period, and of the longer period of its products, which therefore hardly interfere.

THORIUM X

Period of average life—5.35 days.
Period of half-change—3.7 days.
Radiation—α-rays, (β)-rays.
Range of α-rays—5.7 cm. of air.
Nearest chemical analogue—Radium.
Parent—Radio-thorium.
Disintegration product—Thorium emanation.

The disintegration theory was originally proposed to account for the continuous reproduction of thorium X from thorium salts (due, as we now know, to the always present and inseparable product, radio-thorium), and for the production from the thorium X of a whole chain of successive products, commencing with the gaseous emanation. When any thorium compound in solution is precipitated by ammonia, the thorium X, which is an alkaline-earth element like radium, is left in solution, whereas if the precipitant is a carbonate, phosphate, &c., no separation is effected, and the precipitated thorium possesses the same radioactivity and, in solution, the same emanating power as initially. If after precipitation of the thorium with ammonia the filtrate is evaporated to dryness and the ammonium salts volatilised by gentle ignition, a non-volatile residue is obtained, very small in quantity but intensely radioactive, the activity being that of thorium X. If the thorium compound used is, for example, one like thorium nitrate, and has been preserved in the solid state, the emanation is retained, and nearly the full amount of the later products of the "active deposit" thorium B to D will be present. Whereas if a solution of thorium which has been kept for the last two days in an open vessel is used, these products will not be present, at least in their full equilibrium quantity. On precipitation with ammonia the thorium X alone is left in solution, the other products, if present, being precipitated with the thorium hydroxide. The activity of the precipitate thus decays for the first twenty-four hours until the re-generation of thorium X overpowers the decay due to the disappearance of the active deposit. After that time the activity steadily rises and attains equilibrium in three or four weeks. At the minimum the α-rays have about 25 to 30 per cent. of their equilibrium value, whilst the penetrating rays are practically absent, as these come entirely from the "active deposit." The emanating power of a solution is a direct measure of the amount of thorium X, so that if the freshly precipitated thorium hydroxide is free from thorium X, when dissolved and tested for emanating power it will be found to have practically none. Conversely the activity of thorium X, especially after ignition, when the emanation is retained, increases for twenty-four hours and then decays regularly to zero in three or four weeks, according to the exponential law with the period of thorium X.

The emanating power of the solution from which a thorium salt has been precipitated with ammonia is at first the same as that of the original thorium in solution, but this decays exponentially with time to zero, according to the period of thorium X. The emanating power of the thorium hydroxide meanwhile steadily rises from zero to the equilibrium value. If, instead of ammonia, pyridine, fumaric acid, or meta-nitro-benzoic acid is used as the precipitant, thorium B as well as thorium X are left in the solution. As thorium B is by far the longest-lived member of the active deposit group, these reagents are to be recommended when it is desired to prepare quickly thorium hydroxide free from both thorium X and its products. Four precipitations with the last-named reagent at intervals of two hours is stated to effect this, and to give a thorium hydroxide of minimum α-activity, no β-activity and no emanating power, the activity of which then increases regularly without the initial decay characteristic of the ordinary precipitate obtained by ammonia.[64]

Thorium X cannot be separated from radium or mesothorium 1, and its close resemblance to the alkaline-earth elements follows from some careful experiments which have been done on the crystallisation of various salts from saturated solutions containing thorium X. Most salts crystallise leaving the active material in the mother liquor, but the barium salts on crystallisation preserve the same proportion of the thorium X as the mother liquor.[65]

Out of acid solutions thorium X cannot be separated by electrolysis or by the action of metals, such as copper, zinc, nickel. Only the "active deposit" is separated. But out of alkaline solutions all the active substances may be deposited by these means. Thorium X follows von Lerch's rule, and is less easily deposited, or is electrochemically less "noble" than its successive products. It has not been volatilised. The period of thorium X is almost identical with that of the radium emanation, and this is the only case of the kind among the known radioelements. This leads to some curious results in the separation of thorianite, when the radioactivity is studied by means of the γ-rays. Preparations obtained, for example, by precipitating barium sulphate in the solution appear to be of constant activity, when what is really taking place is the simultaneous decay of thorium X and the regeneration of radium emanation, leading to a decay of the γ-rays due to thorium C and a growth due to those of radium C.

THORIUM EMANATION

Period of average life—76 seconds.
Period of half-change—53 seconds.
Radioactive constant—0.0131 (sec.)$^{-1}$.
Radiation—α-rays.
Range of α-rays—5.5 cm. of air (?).
Nearest chemical analogue—Radium emanation.
Parent—Thorium X.
Disintegration product—Thorium A.

The short period of the thorium emanation and, in consequence, its rapid reproduction after its removal from thorium or thorium X preparations by a current of air, are the features that distinguish it chiefly from the radium emanation. Apart from this the two emanations are very similar, both giving α-rays on disintegration, producing active deposits, being condensed at liquid air temperature and absorbed by charcoal at ordinary temperature but not by chemical reagents. The condensation of the thorium emanation under ordinary circumstances is not so sharp as that of the radium emanation, condensation commencing as high as $-120°$, and being complete at $-155°$. Its coefficient of diffusion is similar to that of the radium emanation, being about 0.1, indicating that the molecular weight is high. Direct measurements by the effusion method have given the molecular weight as between 200 and 210.[65A] As already explained, under *Thorium*, the emanation escapes from various solid thorium compounds with very different degrees of facility, the solid nitrate and ignited oxide giving it up least, and the carbonate and hydroxides most readily. As will be explained in the next section, the product of the thorium emanation is an excessively short-lived α-ray product, the effects of which in ordinary work are indistinguishable from that of its parent. Owing to the relatively long period of the other radioactive members of the thorium active deposit, the emanation must be passed for considerable periods through vessels before the active deposit produced on the walls becomes evident. Once produced the activity continues for two days or more. In the case of the radium emanation the converse is the case. Even when stored for only a few seconds in any vessel, the radium emanation leaves behind a notable active deposit of radium A, the activity of which decays with characteristic rapidity after the emanation is withdrawn. Thorium A decays too rapidly to be distinguishable from the emanation in such experiments. The best source of thorium emanation is a preparation of radio-thorium (*q.v.*). The amount of thorium emanation generated by a thorium solution under conditions similar to that for the case of a standard thorium solution can only be used as a measure of the thorium in the case of minerals in equilibrium. For minerals the emanation affords a direct measure of the thorium present, which can be far more quickly performed than a chemical estimation. A steady stream of air is passed through the solution into a brass cylinder, containing an axial electrode in an insulated support, connected to the gold-leaf of a separate electroscope. The air stream may be kept constant by means of a T-tube in the circuit dipping beneath water, by allowing part of the air always to blow off through the water, thus preserving a constant head. The leak in the electroscope, when it has become constant, is then compared with that produced by a similar volume of a standard solution of a thorium mineral under identical conditions.

THORIUM "ACTIVE DEPOSIT"

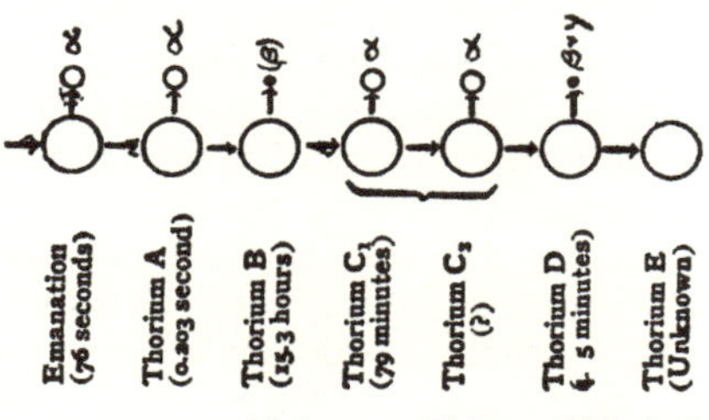

	ThA.	ThB.	ThC_1 & ThC_2.	ThD.
Period of average life . .	0.203 sec.	15.3 hours.	79 mins.	4.5 mins.
Period of half-change . .	0.014 sec.	10.6 hours.	55 mins.	3.1 mins.
Radiation	α-rays.	(β)-rays.	Two α-rays.	β- and γ-rays.
Ranges of α-rays . . .	?	—	5.0 and 8.6 (cm. of air).	—
Penetration power of β-rays (mm of Al for half-absorption)	—	0.05	—	0.441
Penetration power of γ-rays (cm. of lead for half-absorption)	—	—		1.5

The first product of the thorium emanation, thorium A, has only been recently discovered.[66] It had previously been known that the actinium and thorium emanations emitted their α-particles in pairs, as shown by the appearance of double scintillations on the zinc sulphide screen. In the case of actinium the emission appears simultaneous, whereas in the case of thorium a distinct but very short time interval separates the two members of the pair.[67] If, in a dark room, the thorium emanation is allowed to diffuse from a radio-thorium preparation into a small cylindrical vessel carrying an axial electrode coated with zinc sulphide and connected to the negative pole of a battery of 1000 volts, the application of the field is accompanied by the instant brightening of the end of the electrode. On disconnecting the battery the luminosity instantly ceases. Another way of demonstrating the existence of this excessively short-lived first member of the thorium active deposit is to carry an endless wire, negatively charged, through small holes in ebonite stoppers closing a tube containing a source of thorium emanation, and to drive the wire round by a motor. On applying a zinc sulphide screen, the wire is found to be active while the motor is driving it, and the activity along the moving wire is found to fall off the more quickly the slower the wire is driven. By this or similar devices the period can be accurately determined. The half-period, 1/7th of a second, is the shortest known, except for an analogous much more rapidly decaying product of the actinium emanation.[68] Both are entirely analogous to radium A of the radium active deposit. In all ordinary experiments the effects of thorium A add themselves to those of the emanation, and are indistinguishable from them.

The second product of the thorium emanation, thorium B, may be considered practically rayless, the very soft (β)-rays it produces not being at all powerful. In consequence, a negatively charged wire exposed for a short time to a very powerful source of thorium emanation, such as can be got by using radio-thorium, and then withdrawn, possesses practically no activity a second after withdrawal, but in the course of a few hours a powerful activity due to thorium C_1 and C_2 develops. This activity attains a maximum in 220 minutes, and then decays slowly at first, but after 5 hours exponentially to zero with the half-period of 10.6 hours, the decay being practically complete in two or three days. In these changes the α-, β-, and γ-rays vary very nearly uniformly, the period of thorium D, the β- and γ-ray producing body, being too short to enter into consideration. According as the time of exposure to the emanation is increased, the initial rise of the activity becomes less and less marked, until after several days' exposure, when the whole of the products are initially in equilibrium, the existence of thorium B makes itself felt merely by a temporary lag over the first five hours after withdrawal of the decay of the activity. This example illustrates well one point in the theory of successive changes. The course of the curves of activity of the thorium active deposit would be identical in every respect if the periods of thorium B and thorium C were exchanged in the series, that is to say, if the first rayless body had the shorter instead of the longer of the two periods. It is only by separating one of the products in the pure state, and measuring its period of decay, that the choice can be made as to which product has the longer period.[69]

The separation of thorium B from thorium C can be effected in various ways. By heating the active wire for a short time to a red-heat, the thorium B alone distils away without any immediate effect on the activity of the wire, which, however, now decays much more rapidly.[70] The sublimate is at first inactive, but acquires an activity reaching a maximum in four hours, and then decaying with the 10.6 hour half-period. The sublimate, therefore, is thorium B. If the wire is heated above 700°, thorium C also volatilises, but heating for a few minutes to 1000° removes all the thorium B and leaves some of the thorium C, which now decays exponentially with the half-period of 55 minutes.

Thorium C may be removed alone from a solution of the thorium active deposit in acids by absorption with animal charcoal, the thorium B remaining in solution. The same is the case when the solution is shaken up with finely divided nickel or exposed to a nickel plate, or when it is electrolysed, thorium C being always the most readily deposited. Zinc, on the other hand, deposits both thorium B and C from a solution, but not thorium X.

So far thorium C has been referred to as a single product. There is, however, strong evidence that the α-ray change of half-period of 55 minutes of thorium C is followed or accompanied by a very rapid change, in which another α-ray of longer range is expelled.[71] This, so far unisolated, short-lived product is called thorium C_2 whilst its product, which gives β- and γ-rays, is called thorium D. Thorium C_1 and C_2, so far as is yet known, always act together as a single substance. The evidence for the existence of two substances is so far confined to the fact that thorium C gives two sets of α-rays of quite distinct ranges, and it is not known whether these are derived from successive or simultaneous changes. At first it was thought that thorium C_1 and C_2 gave also the β- and γ-rays, but lately a distinct product, thorium D, giving the β- and γ-rays only and having a half-period of only 3.1 minutes, has been separated.[72] By recoil, thorium D is obtained in a pure state from the active deposit of thorium. The powerful α-ray disintegrations of thorium C_1 and C_2, one of which is the most powerful known, causes a considerable proportion of the product, thorium D, to recoil. If in the separation of thorium C by electrolysis, or shaking with nickel, the operation is accomplished in a few seconds, the thorium C is at first free from thorium D. Its α-rays decay exponentially with the 55 minute interval, whilst its β-rays are at first zero, but increase to a maximum with the 3.1 minute half-period of thorium D. Thorium D is more easily soluble in acids and is more volatile (being completely volatilised by heating the active wire for 30 seconds in a Bunsen flame) than any of the other products.[73] In this respect it is an exception to von Lerch's rule already referred to (p. 30). Thorium C_1 and C_2 do not give β- and γ-rays, but these are given entirely by thorium D, which gives no α-rays. Thorium D is the last-known member of the series. The decay of the thorium active deposit is regular and complete, no feeble residual activity remaining as in the case of radium.[74] As to the nature of thorium E, the ultimate product, there is at present no hint. Neither can its atomic weight yet be estimated, for the total number of α-particles expelled per atom of thorium passing through the disintegration series is, unlike the case of the uranium series, still quite unknown. In spite of the recent changes of nomenclature thorium D retains its original name, thorium C_1 and C_2 having previously been termed thorium B and C.

ACTINIUM

Atomic weight—Unknown..
Period of average life—Unknown.
Radiation—Rayless.
Parent—Unknown. Probably uranium is the ultimate parent.
Disintegration product—Radio-actinium.
Nearest chemical analogue—Lanthanum. In basiscity between lanthanum and calcium.

As the above headings indicate, actinium is still very little known. Yet its discovery was made by Debierne, in the iron group separated from pitchblende, very shortly after the discovery of polonium and radium. The total absence of knowledge of its most important radioactive characteristic, its period, which at present is wholly incalculable, alone makes it unique. All that can be said is that the activity of the series appears to be permanent within the few years it has been known, and therefore its period presumably must be at least several times as long as that of polonium. This is perhaps the most important datum in radioactivity still lacking. Next perhaps in importance is the atomic weight, the total absence of knowledge of which, of course, extends to the whole series, comprising six or seven members, of which certainly four, and possibly five, emit α-particles. It is a constant constituent of uranium minerals, so far as these have been examined, and this points to uranium being the ultimate parent, but no production of actinium from uranium or any of its products has as yet been observed. Its peculiarity is its great rarity. The α-activity of uranium minerals is 4.64 times that of the contained uranium. Of this all but 0.28 is contributed by the uranium-radium-polonium series proper, the several activities of the various α-ray products being in agreement with the view that uranium expels two and the five other products one α-ray each, per atom disintegrating. Now four or five members of the actinium series expel α-particles, and yet the total proportion of the α-radiation of uranium minerals contributed by the actinium series is estimated as only 0.28 out of this 4.64. Were the actinium series in the main uranium-radium-polonium series its α-activity should be similar to that of the radium series. It seems most probable that one of the members of the uranium series has two ways of disintegrating, the one mode, in which by far the greater proportion of the atoms disintegrate, giving ultimately polonium and the other giving actinium. In consequence of what has been said, it will be clear that actinium must necessarily be excessively scarce even compared with radium. To produce an actinium preparation, which, when in equilibrium with its products, shall have the activity of a given mass of radium in equilibrium, some ten or twelve times as much mineral must be worked up. Moreover, the final complete concentration of actinium from lanthanum, &c., has not yet been successfully accomplished, so that actinium preparations are at once excessively costly, and at the same time disappointing in the intensity of their activity, as compared with radium.

Actinium itself is rayless, and its first product, radio-actinium, has a period of half-change of 19.5 days. As often separated from minerals, free from radio-actinium, actinium preparations are hardly at all active, but the activity and emanating power increase enormously over a period of several months, a behaviour which is characteristic of the substance.

The whole actinium series bears a remarkably close analogy to that of thorium. The most striking property of actinium preparations is their emanating power, and indeed Giesel gave the name "Emanium" to the substance before the identity of his and Debierne's preparations had been established. Actinium is associated with the rare-earths, and usually is prepared mixed with the hydroxides of these substances. These compounds part with their emanation with great ease even in the solid state. In addition the actinium emanation has an average life-period of only a few seconds, and thus, after removal, very rapidly reforms from its parent. If a preparation of actinium is held over a large zinc sulphide screen in the dark, the emanation streaming away from the preparation illuminates the screen, and the luminosity is blown hither and thither by the slightest current of air. This, at one time, unique property is also shown in even greater degree by suitable preparations of the newer radiothorium. Actinium and its products give powerful α-rays, and preparations 100,000 times as active as uranium oxide in this respect have been described. The β-rays are relatively feebly penetrating. The γ-rays are both feeble in activity and very feebly penetrating, even from very active preparations.

Actinium is easily obtained with iron and rare-earths by precipitating barium as sulphate in the acid solution, and this accounts for its presence in pitchblende residues. In the working up of these, the acid solutions, after removal of polonium with sulphuretted hydrogen, are oxidised and precipitated with ammonia, the actinium being precipitated. The precipitate may be extracted with dilute hydrofluoric acid, the insoluble part, consisting of La, Di, Ce, and Th, retaining most of the actinium. After transformation into chlorides the actinium is precipitated by oxalic acid, and the oxalates, after ignition to oxides, are converted into nitrates, combined with nitrate of magnesium, or of manganese, and fractionally crystallised. The actinium accumulates in the mother liquor. The rare-earth element most closely associated to actinium is lanthanum, and by this latter fractionation, with the double magnesium nitrate, part of the lanthanum may be removed in the crystals free from actinium. Ionium may be separated by precipitation with sodium thiosulphate, in presence of thorium, the actinium remaining in solution. Auer von Welsbach places actinium between lanthanum and calcium in chemical properties.[20] He observed that in the presence of ammonium salts the precipitation of actinium is far from complete, but it is completely precipitated in presence of manganese from basic solutions as a manganate. This reaction he found of great service in the separation of actinium from pitchblende residues on a large scale.

Observations, just published,[74A] on the penetrating rays of actinium, prepared many years previously, have shown a marked decay of the activity during the last three years, amounting to about 10%. This must mean either that the period of average life of the element is only about thirty years, or that, as in the case of thorium, an intermediate product of long life exists between actinium and radio-actinium. The first explanation makes the failure to detect a production of actinium difficult to understand, for this should be easy to do if the period is only thirty years. The second explanation has certainly a very strong analogy to recommend it.

RADIO-ACTINIUM

Period of average life—28.1 days.
Period of half-change—19.5 days.
Radiation—α-rays. (β)-rays.
Range of α-rays—4.55 cm. of air.
Penetration power of (β)*-rays*—Half absorbed by 0.04 mm. of aluminium.
Nearest chemical analogue—Unknown.
Parent—Actinium.
Disintegration product—Actinium X.

Radio-actinium may be separated from its parent and its products by forming in its solution a fine amorphous precipitate, such as sulphur, by adding a little sodium thiosulphate to the strongly acid solution.[75] The actinium and actinium X remain in solution, whilst the radio-actinium is precipitated. The separation, however, is described as rather uncertain. When an actinium preparation is treated with dilute hydrochloric acid, a small part usually is undissolved, and this contains a larger proportion of the radio-actinium than the solution. Freshly prepared, the activity of radio-actinium consists entirely of non-penetrating α- and (β)-rays, which increase, in about three weeks, to a maximum of two or three times the initial value, due to the production of actinium X and its products. The penetrating rays and the emanating power of the preparation increase from zero to a maximum in the same time, and then the total activity decays exponentially with the half-period of about twenty days, the decay being practically complete after four months. Actinium itself is often not completely precipitated by ammonia, whereas radio-actinium appears to be more easily precipitated. By fractional precipitation of an actinium solution with ammonia the radio-actinium is concentrated in the first fractions. The behaviour of the substance is not very definite, and the methods usually successful in effecting the separation sometimes fail. No doubt this uncertainty will disappear on more thorough examination. By analogy with the other series, radio-actinium should be identical chemically with thorium.

ACTINIUM X

Period of average life—14.8 days.
Period of half-change—10.2 days.
Radiation—α-rays.
Range of α-rays—4.17 cm. of air.
Nearest chemical analogue—Radium.
Parent—Radio-actinium.
Disintegration product—Actinium emanation.

Actinium X resembles thorium X completely in chemical nature,[65] and may be separated from actinium or radio-actinium by precipitation with ammonia.[76] The only difference between this separation and that of thorium X is that the "active deposit" of actinium is separated with the actinium X, but in the ignition of the ammonium salts it is volatilised. The result is therefore similar, so far as the ignited preparation is concerned. Initially it is free from "active deposit," and immediately starts to produce it. The α-rays increase in intensity about one-third, and the β-rays grow from zero to a maximum, in three or four hours, and then the total activity and emanating power decay exponentially with the half-period of ten or eleven days. The actinium hydroxide so obtained is, like the thorium hydroxide, precipitated with fumaric or meta-nitro-benzoic acid, initially free from the "active deposit," and in consequence the activity rises from preparation without any initial decay. The α-activity is initially about 28 % of the equilibrium value, which is due entirely to radio-actinium, but owing to the indefiniteness already referred to in the behaviour of the latter substance, it is apt to be separated to some extent from the actinium in the chemical operations for removing actinium X. As in the case of thorium X, the emanating power of a solution is the direct measure of the amount of actinium X present therein at the time of the measurement.

ACTINIUM EMANATION

Period of average life—5.6 seconds.
Period of half-change—3.9 seconds.
Radioactive constant—0.018 (sec.)$^{-1}$.
Radiation—α-rays.
Range of α-rays—5.40 cm. of air.
Nearest chemical analogue—Radium emanation.
Parent—Actinium X.
Disintegration product—Actinium A.

The emanation of actinium, except that it has a still shorter period than that of thorium, resembles the other two emanations closely. Actinium preparations generally consist for the most part of hydroxides of the rare-earths, which evolve their emanation with great facility, and this and the short period account for the beautiful effects due to the emanation obtainable from most actinium preparations, as already described under *Actinium*. The diffusion coefficient is 0.11, distinctly greater than that of the other two emanations.[77] In still air, of course, the emanation can diffuse only a very short distance from its point of origin before it disintegrates. The emanation is condensed by cold over the range from $-100°$ to $-143°$.[78] So far as is known, it is not chemically absorbed by any reagent, and is therefore probably a member of the argon family.

Quite recently it has been found that the change of the actinium emanation is complex. Two sets of α-rays of different ranges are expelled, the α-ray of longer range coming from the product of the emanation. This product has an excessively short period of life, and is attracted to the negative electrode like an active deposit. It has therefore been named Actinium A, and will be considered under the "Active Deposit."

ACTINIUM "ACTIVE DEPOSIT"

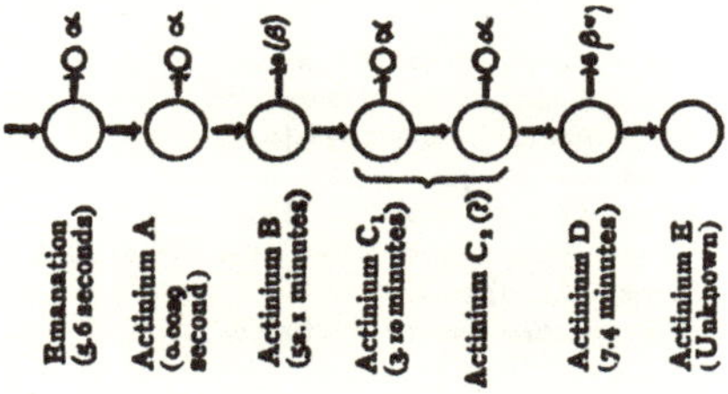

	AcA	AcB	AcC_1 and AcC_2	AcD
Period of average life .	0.0029 sec.	52.1 min.	3.10 min.	7.4 min.
Period of half-change .	0.0020 sec.	36.1 min.	2.15 min.	5.10 min.
Radiation . . .	α-rays	β-rays	α-rays	β- and γ-rays
Range of α-rays . .	6.16 cm.	—	5.12 cm. of air	—
Penetration power of β-rays (mm. of Al for half-absorption)	—	<0.04		0.24
Penetration power of γ-rays (cm. of lead for half-absorption)	—			0.57 (after 0.8 cm.)

The recently discovered first member, actinium A, gives α-rays, and has the shortest period of any disintegration product so far measured. The period of half-change has been measured accurately by a rotating disc method, and found to be exactly 1/500th of a second.[68] Actinium A may be put into evidence as a distinct product of the emanation by methods similar to those described under *Thorium A*. It is of interest that Giesel, so long ago as 1903,[79] working with a powerful emanium preparation observed that the emanation could be influenced and directed by an electric field. If the preparation was contained in an earthed vessel with an opening of any shape, on bringing a zinc sulphide screen, negatively charged, opposite the opening, a bright image of the latter appeared on the screen as a phosphorescent patch, and disappeared instantly the field was interrupted or screened. He termed the phenomenon the "E-Rays." This experiment is very analogous to that already described under *Thorium A*. In all ordinary work actinium A and the emanation act together as a single product, and their effects are superimposed.

The actinium active deposit bears the closest analogy to the thorium active deposit, the periods of the (β)-ray and α-ray producing members being, however, much shorter, and that of the β- and γ-ray producing member somewhat longer than in the latter case. The period of the active deposit as a whole is very similar to, being rather longer than, that of the radium active deposit after the rapid changes due to the A products are at an end. There is a slight initial lag in the decay of the actinium active deposit, similar to, but much less pronounced than, that in the case of the thorium active deposit. After this initial lag the decay continues exponentially to zero with the half period of 36.1 minutes due to actinium B, the longest-lived member of the group. Actinium B is more volatile than actinium C, and on heating a wire or plate coated with the active deposit strongly for half a minute the α-activity is not itself volatilised but decays with great rapidity after the heating, with the period of actinium C, the actinium B being almost completely volatilised. The β-activity of the wire or plate after heating, on the contrary, rapidly increases to a maximum and then decays with the somewhat longer period of actinium D. The latter substance is thus volatilised with the actinium B. It can also be prepared pure by recoil from the active deposit, the active plate being positively charged and opposed by a similar negatively charged plate, at a short distance, on which the recoiled product is received. The yield may be as much as 50 % of the theoretically possible. Its activity is solely confined to β- and γ-rays, and decays exponentially with the half-period of 5.1 minutes. Actinium D is alone withdrawn from an acid solution of the active deposit by boiling with animal charcoal or platinum black.[80]

With regard to the complexity of actinium C, the evidence is exactly of the same kind as in the case of thorium C, but is much less conclusive, because the ranges of the α-rays given by actinium C are extremely similar, though apparently not identical. The decay of the active deposit is complete, and is regular right up to the end, and there is no evidence whatever as to the identity or atomic weight of the ultimate product "actinium E."

POTASSIUM AND RUBIDIUM [81]

A few words in conclusion may be said regarding the radioactivity of these two elements, though it is doubtful at the present time whether we have to do here with cases of atomic disintegration of the type established for the other radio-elements. All potassium salts show a feeble β-activity, about 1/1000th part of the β-activity of uranium. The fact that the so-called "potassium cyanide" of commerce is often exceptional is due to these preparations being largely sodium cyanide. The activity appears to be a specific atomic property of potassium, and has not so far been concentrated from it. Of the other alkali-metals and their salts, lithium, sodium, and cæsium possess no detectable activity. Rubidium, however, possesses a β-activity similar to that of potassium, but distinct from it. The specific activity of the two elements for thick layers of their salts is very similar, but the potassium β-ray is about nine times more penetrating than that of rubidium. Hence the rays come from greater depths in the case of potassium salts, and the activity of rubidium, allowing for the absorption of the radiations of the salt itself, must really be much greater than that of potassium. The potassium β-rays are described as heterogeneous and of varying degrees of penetrating power. The most penetrating are nearly as penetrating as those of uranium X. They are easily deviated by an electric field. On account of their feebleness these cases of radioactivity are difficult to investigate. The alkali-metals under the influence of light emit cathode-rays, that is to say, they possess photo-electric sensitiveness. No special examination seems to have been made of the influence of light on the radioactivity of potassium. It is difficult to believe that the atoms of potassium and rubidium are really disintegrating like those of uranium and thorium until products of the disintegration have been isolated or similar further evidence of the process is obtained. Naturally the facts may be explained by supposing the existence of minute quantities of new elements, not yet separable chemically from potassium and rubidium respectively, as the cause of the phenomena. But of the existence of two new specific types of β-radioactivity it is not possible to doubt.

REFERENCES

[1] Becquerel, Compt. rend., 1896, **122**, 420.
[2] P. Curie, Mme. Curie, and G. Bémont, Compt. rend., July and Dec. 1898.
[3] Debierne, Compt. rend., Oct. 1899 and Apr. 1900.
[4] Rutherford, Phil. Mag., 1900 (v.), **49**, 1.
[5] *Ibid.*, 161.
[6] Rutherford and Soddy, Trans. Chem. Soc., 1902, **81**, 321 and 837; Phil. Mag., 1902 (vi.), **4**, 370 and 569.
[7] *Ibid.*, 1903 (vi.), **5**, 576.
[8] Crookes, Chem. News, 1903, **87**, 241.
[9] Rutherford, Phil. Mag., 1905 (vi.), **10**, 163; 1906 (vi.), **11**, 166; **12**, 134, 348 *et seq.*
[10] Ramsay and Soddy, Proc. Roy. Soc., 1903, **72**, 204 and 346.
[11] Rutherford and Royds, Phil. Mag., 1909 (vi.), **17**, 281.
[12] Bragg, Phil. Mag., 1910 (vi.), **20**, 385.
[13] Geiger, Proc. Roy. Soc., 1910, **83A**, 511.
[14] Geiger and Nuttall, Phil. Mag., 1911 (vi.), **22**, 619.
[15] Kolowrat, Le Radium, 1910, **7**, 1.
[16] Bragg and Kleeman, Phil. Mag., 1904 (vi.), **8**, 719 and 726; 1905 (vi.), **10**, 318 and 600.
[17] Joly and Fletcher, Phil. Mag., 1910 (vi.), **19**, 630.
[18] Regener, Ber. D. physikal. Ges., 1908, 78; Rutherford and Geiger, Proc. Roy. Soc., 1908, **81A**, 141.
[18A] Danysz, Compt. rend., 1911, **153**, 339.
[19] Rutherford and Geiger, Phil. Mag., 1911 (vi.), **22**, 621.
[20] Auer von Welsbach, Sitzungsber. K. Akad. Wiss. Wien, 1910, **119** (ii.a), 1.
[21] Soddy, Trans. Chem. Soc., 1911, **99**, 72.
[22] Lloyd, J. Physical Chem., 1910, **14**, 476.
[23] Hahn and Meitner, Physikal. Zeitsch., 1910, **11**, 493.
[24] Haitinger and Ulrich, Sitzungsber. K. Akad. Wiss. Wien, 1908, 117, (ii.a), 619.
[25] McCoy, Phil. Mag., 1906 (vi.), **11**, 176.
[26] Soddy, Phil. Mag., 1909 (vi.), **18**, 858; 1910 (vi.), **20**, 342.
[27] Boltwood, Amer. J. Sci., 1908 (iv.), **25**, 269.
[28] Bronson, Phil. Mag., 1908 (vi.), **16**, 291; Geiger and Marsden, Physikal. Zeitsch., 1910, **11**, 7.
[29] Soddy and Russell, Phil. Mag., 1909 (vi.), **18**, 620.
[30] Lloyd, J. Physical Chem., 1910, **14**, 509.
[31] Soddy, Phil. Mag., 1909 (vi.), **18**, 861.
[32] Ritzel, Zeitsch. physikal. Chem., 1909, **67**, 725.
[33] Keetman, Jahr. Radioaktivität, 1909, **6**, 269.
[34] *Compare* Antonoff, Phil. Mag., 1911 (vi.), **22**, 431.
[35] Soddy, Phil. Mag., 1910 (vi.), **20**, 342.
[36] Boltwood, Amer. J. Sci., 1906, **22**, 537; 1907, **24**, 370.
[37] Rutherford, Phil. Mag., 1907 (vi.), **14**, 733.
[38] Soddy, Le Radium, 1910, **7**, 295.
[39] Mlle. Gleditsch, Le Radium, 1911, **8**, 256.
[40] De Forcrand, Compt. rend., 1911, **152**, 66.
[41] *Compare* Soddy, Phil. Mag., 1909, **18**, 846.

[42] Rutherford, Radioactivity, 2nd edition, 468.
[43] Rutherford, Phil. Mag., 1908 (vi.), **16**, 300; Gray and Ramsay, Trans. Chem. Soc., 1909, **95**, 1073; Debierne, Compt. rend., 1909, **148**, 1264.
[44] Royds, Phil. Mag., 1909 (vi.), **17**, 202; Watson, Proc. Roy. Soc., 1909, **83A**, 50.
[45] Debierne, Compt. rend., 1910, **150**, 1740.
[46] Gray and Ramsay, Proc. Roy. Soc., 1911, **84A**, 536.
[47] Gray and Ramsay, Trans. Chem. Soc., 1909, **95**, 1073; Rutherford, Phil. Mag., 1909 (vi.), **17**, 723.
[48] Ramsay and Soddy, Proc. Roy. Soc., 1903, **72A**, 204; **73A**, 346.
[49] Mlle. Ramstedt, Le Radium, 1911, **8**, 253.
[50] P. Curie and Danne, Compt. rend., 1904, **138**, 748.
[51] Makower, Le Radium, 1909, **6**, 50.
[52] von Lerch, Ann. Physik, 1906 (iv.), **20**, 345.
[53] Hahn and Meitner, Ber. D. physikal. Ges., 1909, **11**, 55.
[54] Hahn and Meitner, Physikal. Zeitsch., 1909, **10**, 697; Fajans, *ibid.*, 1911, **12**, 369.
[55] Antonoff, Phil. Mag., 1910 (vi.), **19**, 825.
[56] Meyer and von Schweidler, Wien Sitzungsber., 1906, **115** (ii.a), 697.
[57] Marckwald, Ber. D. chem. Ges., 1902, **35**, 2285 and 4239; 1903, **36**, 2662.
[58] Mme. Curie and Debierne, Compt. rend., 1910, **150**, 386.
[58A] Koppel and Holtkamp, Zeitsch. anorg. Chem., 1910, **67**, 266.
[59] Hahn, Ber. D. chem. Ges., 1907, **40**, 1462.
[60] Marckwald, Ber. D. chem. Ges, 1910, **43**, 3420; Soddy, Trans. Chem. Soc., 1911, **99**, 72.
[61] Hahn, Chem. Zeit., 1911, **34**, 845.
[62] Hahn, Physikal. Zeitsch., 1908, **9**, 246.
[63] Russell and Soddy, Phil. Mag., 1911 (vi.), **21**, 130.
[63A] Mlle. Leslie, Le Radium, 1911, **8**, 356.
[64] Schlundt and Moore, J. Physical Chem., 1905, **9**, 682.
[65] Strömholm and Svedberg, Zeitsch. anorg. Chem., 1909, **61**, 338; **63**, 197.
[65A] Mlle. Leslie, Compt. rend., 1911, **153**, 328.
[66] Rutherford and Geiger, Phil. Mag., 1911 (vi.), **22**, 621.
[67] Geiger and Marsden, Physikal. Zeitsch., 1910, **11**, 7; Geiger, Phil. Mag., 1911 (vi.), **22**, 201.
[68] Moseley and Fajans, Phil. Mag., 1911 (vi.), **22**, 629.
[69] Rutherford, Radioactivity, 2nd edition, 351.
[70] Miss Slater, Phil. Mag., 1905 (vi.), **9**, 628.
[71] Hahn, Phil. Mag., 1906 (vi.), **11**, 793.
[72] Hahn and Meitner, Ber. D. physikal. Ges., 1909, **11**, 55.
[73] von Lerch and von Wartberg, Sitzungsber. K. Akad. Wiss. Wien, 1909, **118** (ii.a), 1575.
[74] von Lerch, *ibid.*, 1907, **116** (ii.a), 1443.
[74A] Mme. Curie, Le Radium, 1911, **8**, 353.
[75] Hahn, Ber. D. chem. Ges., 1906, **39**, 1605; Physikal. Zeitsch., 1906, **7**, 855.
[76] Godlewski, Phil. Mag., 1905 (vi.), **10**, 44.
[77] Bruhat, Compt. rend., 1909, **148**, 629; Le Radium, 1909, **6**, 67; Russ, Phil. Mag., 1909 (vi.), **17**, 412.
[78] Kinoshita, Phil. Mag., 1908 (vi.), **16**, 121.
[79] Giesel, Ber. D. chem. Ges., 1903, **36**, 342.
[80] Hahn and Meitner, Physikal. Zeitsch., 1908, **9**, 649.
[81] Campbell and Wood, Proc. Camb. Phil. Soc., 1907, **14**, 15; Campbell, *ibid.*, 211; 1909, **15**, 11.

INDEX

Made in the USA
Middletown, DE
05 November 2026

36544998R10058